Flight Paths to Success: Career Insights from Women Leaders in Aerospace

Rhonda Walthall, Editor and
Brenda Mitchell, Editor

400 Commonwealth Drive
Warrendale, PA 15096-0001 USA
E-mail: CustomerService@sae.org
Phone: 877-606-7323 (inside USA and Canada)
724-776-4970 (outside USA)
FAX: 724-776-0790

Library of Congress Catalog Number 2021931770
http://dx.doi.org/10.4271/ 9781468603033

Information contained in this work has been obtained by SAE International from sources believed to be reliable. However, neither SAE International nor its authors guarantee the accuracy or completeness of any information published herein and neither SAE International nor its authors shall be responsible for any errors, omissions, or damages arising out of use of this information. This work is published with the understanding that SAE International and its authors are supplying information but are not attempting to render engineering or other professional services. If such services are required, the assistance of an appropriate professional should be sought.

ISBN-Print 978-1-4686-0256-2
ISBN-PDF 978-1-4686-0303-3
ISBN-ePub 978-1-4686-0304-0

To purchase bulk quantities, please contact: SAE Customer Service

E-mail: CustomerService@sae.org
Phone: 877-606-7323 (inside USA and Canada)
724-776-4970 (outside USA)
Fax: 724-776-0790

Visit the SAE International Bookstore at books.sae.org

Chief Growth Officer
Frank Menchaca

Publisher
Sherry Dickinson Nigam

Development Editor
Dan Reilly

Director of Content Management
Kelli Zilko

Production and Manufacturing Associate
Erin Mendicino

Contents

CHAPTER 7

Charmaine Chin   39

CHAPTER 8

Nagin Cox   43

CHAPTER 9

Kirsten Dreggors   47

CHAPTER 14

Linda Flores   77

CHAPTER 15

Mary Lee Gambone   81

CHAPTER 16

Neri Ganzarski   87

CHAPTER 21

Miyuki Humer 119

CHAPTER 22

Jamie Korman 123

CHAPTER 23

Lorraine Martin 129

CHAPTER 28
Tammy Reeve — 165

CHAPTER 29
Tamaira Ross — 169

CHAPTER 30
Ginger Shao — 175

> " Leaders tend to be passionate about what they do. I urge you to do whatever it takes to get yourself in a position where you love what you do, where you care about what you do, where you want to inspire others, build great things, go great deeds. Only then will you find true fulfillment. Only then will you be a brilliant leader. "
>
> *Shelly Lazarus, Chairman Emeritus,*
> *Ogilvy and Mather Worldwide*

Quote from "*Letters from Leaders. Personal Advice for Tomorrow's Leaders from the World's Most Influential People*," compiled by Henry O. Dorman, The Lyons Press, 2009, ISBN 978-59921-501-3.

Requests for permission to reproduce: The Globe Pequot Press, Attn: Rights and Permissions Department, PO Box 480, Guilford, CT 06437.

Foreword

I grew up in an aviation family. My father, Frank Piasecki, was one of America's prominent helicopter pioneers, and aviation was the first language we spoke at home. He founded his own trailblazing helicopter engineering and manufacturing company and insisted that all his children learn the business. On Saturdays he would take us into the office with him, starting at age 5 or so. Naturally, we all learned about the wonders of aviation including the smell of Jet A, the beauty of runway lights, and the rhythm of airplanes returning to land as we were perched on my father's test stands at the end of our "work day" on the edge of Philadelphia International Airport. But more importantly, we learned just how much hard work, disciplined and innovative thinking, drive, and dedication are involved in this business. This early learning has served as a solid foundation for my perspective as a woman in the aerospace industry.

When I got to college, my engineering classes were taught by male professors and often consisted of about 85 percent male students. After earning my engineering degree I joined United Technologies, and after earning my MBA from Wharton I joined The Boeing Company. Both were obviously male-dominated companies competing in a male-dominated industry.

At that time, naively, I didn't focus on how my gender might limit or define my contribution. I focused on working hard, performing at the highest level I could, and making my absolute best contribution. My goal was to have a significant positive impact on the business in my current assignment while always keeping an eye out for the next opportunity.

Aerospace is a tough business, and challenges abound. A collaborative spirit, an insatiable appetite for learning, and a relentless work ethic go a long, long way. I've also benefited immensely by learning from and working for people I admire and respect, both men and women. Such role models have given me opportunities to succeed and (equally important) opportunities to fail, and have helped me discover depths of potential I didn't know I had. I have also been very fortunate to have a profoundly supportive husband, six siblings, and three wonderful sons. We work closely together as a family to keep work and life balanced.

Many of my strongest career advocates in aerospace have been men. But of course, like all women in aerospace, I've encountered my share of gender bias too. I was once informed that despite being the best-qualified candidate for a sales position I sought, I was ruled out because the vice president leading the organization thought I would inevitably start a family. Another opportunity was derailed on the basis that I was too "frou frou." And I was actually told by my assigned mentor that I was unsuited for yet another sales position I coveted because it involved dealing with one of the best negotiators in the industry, so (as a woman) I'd be in way over my head. He sternly counseled me not to apply, but I did anyway and landed the job. The next year I sold more airplanes in the first few years than any of my counterparts.

Like many of my sister colleagues in this book, I've done my best to not just work hard but work harder than my male counterparts. I've understood that it's not enough to match their performance—I must try to outperform them. I've approached my work with the attitude that I need to be smarter, more collaborative, better prepared, more resourceful, and tougher than the guys. In the end, this attitude has profoundly enriched my work and life because I have the knowledge that I've consciously, energetically given and done my best.

We all define career success individually. But it's also essential that women engage and support one another. Madeline Albright once said, "There's a special place in hell for women who don't help other women," and I agree. One of the most valuable gifts we can share with each other as women in this challenging business is our experience. I've always found comfort in the knowledge that there are others like me on this challenging journey, others who share my commitment to this extraordinary endeavor and who understand the unique pressures we face as women in the aerospace industry. I've found insight and affirmation in these kindred personal experiences and, sometimes, the needed faith and strength to persist and strive further.

This book is a remarkable compendium of successful women in aerospace and their individual journeys. I'm confident that you will find their stories as instructive and inspiring as I do.

Nicole Piasecki

Nicole Piasecki is currently Chairman of Seattle University's Board of Trustees.

Piasecki retired from Boeing as Vice President and General Manager of the Propulsion Systems Division of Boeing Commercial Airplanes. She served in a variety of leadership positions during her 23-year career at Boeing Commercial Airplanes, most recently as vice president of Business Development and Strategic Integration, starting in January 2010. Prior to that Piasecki held positions as president of Boeing Japan, Senior Vice President of Business Strategy and Marketing for Commercial Airplanes, and Vice President for Sales for Leasing Company customers. Piasecki played an instrumental role in the investment in the Boeing product line including the launch of the 787 Dreamliner and 737 MAX.

She earned a bachelor's degree in Mechanical Engineering from Yale University and a Master of Business Administration degree in operations management from The Wharton School at the University of Pennsylvania, which included studies at the Keio Business School in Japan.

Piasecki has served in several government capacities during her career. In 2003, Transportation Secretary Norman Mineta named her to the Federal Aviation Administration's Management Advisory Council. Starting in 2010, she served a one-year term as Chair of the Safety Subcommittee of the Future of Aviation Advisory Committee. In 2012, she was appointed to the board of directors of the Seattle Branch of the Federal Reserve Bank of San Francisco.

Piasecki is a director at Weyerhaeuser Co., British Aerospace Systems PLC, Howmet Aerospace, and the Chairman of Seattle University's Board of Trustees. She is also an advisor to Mitsubishi Heavy Industries in Tokyo.

Piasecki has a single-engine pilot's license, is married, and has three sons.

Preface

The aerospace industry is awesome! We are always at the forefront of technology—we do really cool stuff! What isn't always at the forefront is women leaders.

The aerospace industry is still a very male-dominated space in 2020. Women engineers make up approximately 10% of the engineering workforce, and very few with STEM degrees reach top levels of leadership. Women are often overlooked for promotion because they lack "the relevant experience" only to see less-qualified male colleagues get the role. How does this happen? What did we miss?

It's not enough for a woman to do a great job and be the best at what she does. The women highlighted in this book have found ways to navigate this industry and achieve success. Our goal was to profile women that have engineering degrees who have achieved significant career success in hopes of recruiting more women to our field or to help guide those that may be experiencing career frustration. We have all been there—there is hope. The common theme among many of these stories is the importance of mentorship and more importantly "sponsorship" and the importance of family support when making difficult career decisions.

The book profiles the personal journeys of 33 women who have been, and continue to be, successful in aviation, space, and academia. Each woman was asked to select one question of several questions in five categories: personal career insight, work-life balance, mentorship/sponsorship, avoiding a career stall, and powering through challenging situations. Each woman shared her unique experiences about work-life integration, resilience, career changes, relocation, continuing education, and career advancement. While reading their stories, we saw that there were many flight paths to success and each women navigated her own way by charting her own course and committing to it. Their stories were published as they wrote them—in their own words.

We started out by thinking we didn't know this many incredible women, but when we worked our networks, this book was the result. Whether you are just thinking about majoring in engineering or are in some phase of your aerospace career, the advice and wisdom presented herein is invaluable, and we wish we had been the beneficiaries of this at the start of our careers. We hope that you will find inspiration and insights to help you chart your career path with less turbulence than those who have navigated this space for the past seven decades.

The sky's the limit!

Brenda and Rhonda

Acknowledgments

We would like to thank the SAE publishing team and especially Sherry Nigam and Dan Reilly for making this book possible. Without Sherry's publishing expertise and Dan's superb organizational support, this would have been a very daunting task. Their professionalism made this a very enjoyable journey.

Secondly, we owe a profound debt of gratitude to the women who chose to participate in this project. They invested a significant amount of time reflecting on and writing their stories, which are as phenomenal as they are.

In addition, Rhonda would like to thank her husband Larry for his patience while countless evenings and weekend hours were spent working on SAE activities.

Finally, we thank you, our readers, and hope you find this book invaluable as you think about your own careers and chart your flight path.

About the Editors

Rhonda Walthall
Technical Fellow
Collins Aerospace

- Email: rhonda.walthall@collins.com
- https://www.linkedin.com/in/rhonda-walthall-20096a18/
- Topics: aircraft health management, prognostics, and diagnostics

Education

- MBA, Pepperdine University, 1993
- BS, Purdue University, Aeronautical and Astronautical Engineering, 1986

Rhonda Walthall is a leader in integrated aircraft health management with 33 years in the aerospace industry. As a Fellow at Collins Aerospace Systems, a division of Raytheon Technologies Corporation (RTX), she focuses on Design for Prognostics and Health Management (PHM) and Model-based Digital Thread initiatives. For the past 16 years, she has been actively engaged in the development of standards and best practices for PHM and Integrated Aircraft Health Management.

Rhonda started her career in 1987 as an Engine Performance Engineer and then a Flight Test Engineer for the McDonnell Douglas Aircraft Company. She joined Northwest Airlines in 1995 as an Engine Condition Monitoring Engineer, where she helped to launch the Flight Operations Quality Assurance program still in use today at Delta Airlines. She joined RTX in 2003 as a Systems Engineer, where she led the development of the Aircraft Systems Health Management aftermarket service offering. She holds four PHM-related patents.

Rhonda is a Fellow of the PHM Society, the former Vice President, and member of the Board of Directors. She is a member of the SAE International Board of Directors and the Audit and Risk Committee. She has been engaged in SAE standards development and leadership since 2005 and has contributed significantly to numerous SAE publications.

Rhonda is an advocate for women in Aerospace and is a 23-year member of Women in Aviation International. She mentors young engineers at work and at Purdue University through her role in the Industrial Advisory Council to the School of Aeronautics and Astronautics. She is a 15-year member of Toastmasters International, earning the Distinguished Toastmaster Award twice and mentoring dozens of fellow members to achieve their communication and leadership goals.

Brenda Mitchell
Aerospace Executive

- Email: brenda.mitchell@azureaerogroupllc.com
- http://linkedin.com/in/brendachristiansonmitchell

Education

- MBA, Boston University, 2002
- Bachelors of Civil Engineering, University of Minnesota,1985

Brenda Mitchell is the Vice President of Engineering, Certification, and ODA (Organization Delegation Authorization) at ALOFT AeroArchitects in Georgetown, DE. Brenda joined ALOFT in 2019 and leads a team of engineers and service professionals who design and certify aircraft systems and interior modifications of VIP, VVIP, and Head of State aircraft. Brenda earned her Bachelor of Civil Engineering degree from the University of Minnesota and her Master's degree in Business Administration from Boston University.

Brenda started her aviation career in 1986 as an engine performance engineer at GE Aircraft Engines. She joined Northwest Airlines in 1990 as a powerplant engineer and quickly focused on developing analytics to drive and improve Engine Condition Monitoring and improved engine reliability. In 1996, Brenda joined the Boeing Company initially in the propulsion organization working 737-700 certification and then moved to the marketing organization. She joined Pratt and Whitney's program management team in 1999 as a program manager for the PW4000 engine family. In 2004, she worked in the Military Engines division primarily focused on aftermarket support and logistics programs where she became an executive director of the Military Engines Customer Support and Services organization, responsible for supporting all of Pratt's fielded military engines.

Since 2016, Brenda has served as a consultant focused on operational improvement to a variety of aerospace entities. She served as a board advisor to the Board of Directors (BOD) of a Civil Engineering consultancy dealing with topics ranging from Merger & Acquisition advisory, CEO succession planning and selection, implementing Lean, and employee engagement. During her advisory tenure, the BOD selected the first two female shareholders.

Brenda is an advocate for women in Aerospace and has mentored many of the females in her organizations. She has a passion for giving back and helping the next generation of female engineers succeed by sharing lessons learned from a 30-year career in Aerospace.

Julie-Ellen Acosta

Vice President (Retired)
Boeing

About the Author

Julie-Ellen retired from the Boeing Company in February of 2017 after 31 years. In August of 2019, she began consulting and providing executive coaching services in the aerospace business sector. Her 30-plus-year career with Boeing includes executive leadership positions in Engineering, Manufacturing, Quality, Program Management, Research and Development, and Human Resources.

Julie-Ellen retired as the Vice President of Manufacturing Excellence for Defense, Space, and Security (BDS) at Boeing. Her responsibilities in the role included leading the development, deployment, and step-function improvement of a common BDS production system across all BDS development and production programs. She has held the role of Vice President of Operations and Quality for the Phantom Works business unit of BDS. Acosta led an integrated team responsible for developing advanced operations and quality integrated design build solutions for Phantom Works around the world.

Her diverse leadership experience during her tenure at Boeing includes serving as vice president of Human Resources for Boeing Commercial Airplanes (BCA). Acosta oversaw the successful negotiation of two major union contracts and the transformation of employee benefits. In addition, as vice president of Leadership Talent Management, she defined and executed a new, comprehensive strategic framework for leadership development and talent management within the Boeing company. Prior to stepping into the special assignment in Human Resources, Julie-Ellen served as the Vice President of Structural and Manufacturing Technologies, Prototyping, and Quality for Boeing Phantom Works, now known as Boeing Research and Technology.

Prior to the vice president roles, Acosta held senior leadership positions for the structural design of the Next-Generation 737 fuselage and program management of the 757 fuselage at Boeing's facility in Wichita, Kan. She also served as Quality and Process Improvement director for the Wichita site. Acosta joined Boeing in 1985 on the defense side of the business as a systems engineer and held several lead engineering positions on both development & production aerospace programs. Prior to Boeing, Julie-Ellen worked in Academia at Wichita State University as the Assistant to the Dean of Engineering and then took on a Design Engineering role with Gates Learjet in Wichita.

Acosta has Bachelor's and Master's degrees in Electrical Engineering from Wichita State University, where she also completed postgraduate studies in structures. Acosta also completed the Executive Management Program at Harvard Business School. Julie-Ellen has four grown children and has five grandchildren.

Julie-Ellen continues to be actively engaged at Wichita State University and serves on several senior advisory boards at the school including the National Advisory Council (NAC), Chairing the College of Engineering Industrial Advisory Board and being a member of the WSU Foundation Investment Committee. She lectures in the College of Engineering on Engineering Leadership. Acosta is a Senior Member of the Institute of Electrical and Electronics Engineers, and a life member of both the American Institute of Aeronautics and Astronautics and the Society of Women Engineers.

Questions and Answers

Personal Career Insight

How did you decide between a leadership vs technical career track?

I spent the first 10 years of my career focused on building technical capability and know-how in the aerospace business. Given I was one of the first female engineers at the first two companies I joined, I knew it was going to be important to build credibility and not immediately jump into management. With my experience in leadership activities during college and having multiple work experiences before graduating, I knew that there would be a strong desire on the part of others to move me to management roles. I wanted to be a lead engineer and go through several projects that would focus on earning experience in the life-cycle of the product. I was fortunate that the first industry leader I worked for saw to it that I spent the time going through both ground and flight school in support of becoming a Controls Systems Specialist within the business.

The experiences were totally worth it and did provide the needed confidence in myself along with earning credibility prior to being formally promoted to a formal management role. I have zero regrets about going through this phase, and it has served me well throughout my entire career. The desire to earn credibility and learn as much as I could about the overall Aerospace business stuck with me as I progressed through the management ranks. The overarching personal goal was to obtain experience in aerospace (design/built/support/certify) so as to build a deep knowledge in the business prior to taking on larger roles such as a Product Line Leader or moving across business functions within the broader business later in my career.

Understanding the pieces of the "whole" became my approach to truly understanding the processes of the product and of the business. This approach provided a solid foundation for moving up the leadership chain and taking on a variety of assignments that crossed functional boundaries.

Work/Life Balance

Did you ever have to make a move to advance your career (within your company or changing companies) that impacted your family life, and how did you balance the two?

I have been asked many times throughout my career to move across both functional and geographical boundaries. All of the moves were associated with gaining a broader perspective of the business and to take on more difficult roles that built upon prior experience. I am the mother of four and I made a local move early in my career to a large company, (Boeing) for the explicit purpose of having opportunities to move laterally, should the opportunity arise.

Early in my career and as a young mother, I wanted to stick close to home while the kids were young. I had a wonderful childhood and my mom was truly my role model. I wanted the same for my kids. I found out at a very young age that I needed a strong support network to help enable the ability to continue in college and then in industry. Staying local to Wichita was the key to this success. I was fortunate that there were multiple opportunities in Wichita for someone with my engineering background, and I felt comfortable making this decision early on in my career. Later in my career, Boeing had a major merger with several other Aerospace companies, and this action drove a change to my career strategy and subsequent opportunities. I knew turning down assignments would eventually limit my career so I had to prepare for the inevitable.

Julie-Ellen Acosta

Each assignment I took on was discussed with my family to ensure they were on board with the new responsibilities. As I transitioned into different assignments with increasing business responsibilities, I actually felt I had more flexibility with my personal life. As I moved from Engineering to Operations, my starting time changed to 5:30 am (or earlier), and as a product leader, my time expanded to be more 24/7. This sounds ominous, but with the new responsibilities did come an opportunity to flex my time allowing me to participate in the kid's activities. I'm often asked about work/life balance (because people would see me working all hours of the day), I can honestly say that it's a highly customized decision for each individual and must be aligned with both your personal and professional needs. Eventually, I had responsibilities not only in the US but also around the globe. I had to learn how to use the clock and I had to have flexibility in my work so I could find balance with being a mom and an effective leader within the Aerospace business.

Julie-Ellen Acosta

Ultimately, I was asked in the late 1990s as to what it would take for me to physically move to a larger part of the business. I was asked about timing and what would work with my family responsibilities. I was actually asked to write a "white paper" describing the scenario. I had the discussion with my family and we defined the timing together. My husband is also an engineer in the Aerospace business, and with us having four kids, we all had to be aligned in our thinking. We all had to be prepared to say "yes" to the move out of Wichita. It was a challenge but ultimately the move became a game changer for my career and for us as a family. We made four moves while I was a Vice President. Three of the four were with the two younger kids.

We made the decision to stay for a full four years in St Louis so the two younger ones could graduate from High School. With the expansion in responsibilities brought the need to strengthen the family being more and more integrated in supporting and being involved in the decision- making process. I am forever indebted to my husband, Carlos and to the four kids for being so supportive. I simply would not and could not have made the career transitions without them. Bottom line work/life balance is highly tailored to the individual, the family, and the specific situation. The strongest advice I can give is to talk to your family about any career decisions and moves. Do not make the mistake of making these life-important decisions in a vacuum. Your family will be with you forever, your "job" will not.

Mentorship/Sponsorship

How important was mentorship/sponsorship for your career? Have they been men or women? How was the relationship established?

Mentorship and Sponsorship have played an important role throughout my entire academic and professional career. Mentorship started with my parents and teachers while in junior high and high school and continued through college and throughout my industry career. During the early years, my parents provided a solid foundation and set guidelines around expectations. Going to college, becoming all that I could become, was always expected. Throughout my academic and industry career, my mentors were both men and women, but during early college courses, all of my mentors were men. In each case, my mentors took it upon themselves to reach out to me and to provide guidance on my Engineering degree and ultimately my career. That is a huge lesson learned and it's an approach I have used in reaching out to others my entire career to help others.

I personally made the decision to become an engineer during my teenage years. I was solid in my conviction to be an engineer. As this became known, both teachers and professionals reached out. I wasn't overly impressed with guidance counselors in the 1970s because the one I had wasn't supportive at all. Thank goodness for my math and science teachers! Mentoring wasn't something I sought, it was the gift of time from those who felt I had the aptitude and who felt I needed the encouragement. I will forever be grateful.

As I moved into the engineering curriculum in college, several professors reached out to provide guidance and advice from day one. I did not experience any kind of bias at Wichita State University (WSU) of any kind from any of my professors. If you had the aptitude, worked hard, asked questions, they would take the time to help and encourage. As I moved towards graduation in college, the Dean of the College of Engineering at WSU reached out to me during the last semester of my BSEE degree and talked to me about staying at WSU versus going out into industry. I took on an administration role, working for the Dean, which included being the Assistant to the Dean along with being an Assistant Professor (freshman studies and Intro to Engineering curriculum), within the College of Engineering. The decision to continue to build my engineering foundation by staying at WSU became critical to solidifying my confidence as an engineer along with giving me an alternative career path while I was raising my first two kids. Along with the administrative role, my contract indicated I would pursue a master's degree in Electrical Engineering. At this point, the Electrical Engineering Department Chair along with the various EE professors stepped in and guided my studies through a series of coursework and research in areas that became my signature work in Aerospace. There was no easy path and the expectation was I'd step up, focus, and be at the top of my studies.

I had not seen the academia opportunity in my future at all. The Dean reaching out in a mentoring capacity to explore the opportunity was a critical inflection point in my career. It was during this time I had the opportunity of a lifetime to work with Duane and Velma Wallace, as well as meet and work with other Aerospace pioneers. The opportunity at Wichita State gave my career a solid foundation in engineering and provided the future lift given I gained a master's degree along with some serious experience in academia and associated aerospace research and teaching.

Throughout my entire student and professional career, I have had mentors work with me on a "preferred" set of experiences that ultimately provided incredible learning and "know-how" in a very complicated aerospace business. I never took the easy way out or took on "side assignments." The focus has been on "line" assignment, which put me in a position to be responsible for product and people. I particularly focused on roles and projects involving "designing" and then transitioned to the "building" of the design. I was fortunate that the mentors I worked with believed in staying in roles for around two to two-and-a-half years. My goal was to leave the assignment with all believing that I was a native of that group. My mentors worked hard to ensure that each assignment built upon the previous assignment. This approach to "managing my talent" became the outline for what would become the Talent Management Planning process for Boeing. As one of my latter assignments as a Vice President, when I took on a role as the VP of "Leadership Development," I was able to look back over the career counseling and mentoring I had had and look at the team of vice presidents across our company and realize we shared the experience of how our talent was developed through a series of intentional assignments and mentoring.

There are three major elements that influence a career: education/training, experiences, and mentoring/sponsorship. I was totally fortunate that the mentors that had reached out and worked with me for over 30 years knew that this "formula" worked, and they took the time to work with me.

Avoiding a Stall

How important is an advanced degree?

The advanced degree in Electrical Engineering and then the postgraduate work in Structures, simply put, differentiated my career. Early in my career, it was necessary for me to be seen as a capable engineer with relevant experiences and know-how in every single job I took on. The master's degree along with the advanced studies allowed me to gain confidence in specific areas of expertise that ultimately led to unique assignments and positions. Confidence and know-how would not have happened without the advanced degree and advanced areas of focus.

As a parent, and one that didn't want to move around a lot, I was focused on creating a wide and lateral set of opportunities in my hometown so that I didn't have to move to gain those experiences. There isn't anything I don't love about being an engineer or being a technical leader. Having a focus on building a wide breadth of skill and capability that allowed me to move to lateral assignments, cross functional boundaries, and experience the broader knowledge of the company and product was an early goal. This approach has always been at the forefront of my decisions around education and experience. I did not know how high I would go in management ,but I did know that I wanted to stay engaged with products and people. Having assignments in every part of the product cycle (research, design, manufacturing, support, supplier management, quality, etc.) was absolutely key in personally staying engaged in the aerospace business and interested throughout a 40-plus-year career.

Powering On

Was there a significant event that changed your career trajectory and what was it?

I have had four major inflection points in my career. The first was to make the decision to stay in academia and pursue advanced degrees and education. The second was to expand my knowledge and know-how in aerospace (Electrical/Avionics and then into Structures). The third was to transition from Engineering into Operations (Manufacturing/Quality) and then a transition back into Research and Development associated with a focus on Structures and Operations, and finally, a fourth transition into Leadership Development and Talent Management at a corporate level.

All four of these inflection points required solid knowledge and know-how of the previous areas. All transitions were leading up to the one assignment that changed the course of my career and also became my passion for the duration of my life.

In 2006 I was asked to interview for the VP of Leadership Development at Boeing. We had just hired a new CEO and he had a different Vision for Developing Leaders at our Company. This assignment appeared to be completely out of my "wheelhouse" other than I had deep knowledge and know-how in the "guts of the business." Picking a leader that had deep technical aerospace business experience was what our CEO and leader of HR were looking for at this specific time for our company. This assignment necessitated a move out of the technical side of the business into Human Resources. This was not a move I had seen coming.

Truthfully, I am glad I made this transition as a Vice President, and I do not believe that it would have worked as well had I done this earlier in my career. One simply had to have the credibility with other company leaders to pull this off. I will have to admit I felt like I was violating all of my previous convictions about being a highly technical engineering leader, but as I got into the assignment, I realized that I could really be a part of effecting a big change as to how we guided and what we expected out of our Company leaders. This also helped me to see that this was a huge opportunity to "give back/pay it forward," whatever phrase you want to use. I wanted the chance to work with other company leaders to surface raw talent and put in processes across the company that would forever help identify talent deep in the pipeline for years and years to come.

I spent four years in this role. This role necessitated working with our Executive Council in defining the role of the Boeing leader, reframing our Leadership Development Curriculum, and ultimately connecting our Performance Management and Succession Planning Processes to the expectations we had for the Boeing Leaders. I was able to work with my peers across the company to put the strategy, the execution plan in place. We defined our "Leadership Development and Talent Management" Processes for the company and were able to identify diverse talent deep in the Talent Pipelines as a result. You might say we "re-engineered" the processes at the Company and Business unit level. This continues to be an ongoing effort, and I am really proud to say that this was a very satisfying role.

As a senior leader at a company, you are going to get asked to take on assignments that may feel like they are not in your "wheelhouse." However, keep an open mind and look at this type of assignment as a broadening opportunity. I went on to lead Human Resources at Boeing Commercial Airplanes. I never saw myself doing something like this in my career. But as someone who has spent a significant amount of their time in both Engineering and Manufacturing, the role was perfect at that point in my career. The other leaders saw themselves in me and I was able to be a true employee advocate as well as a steward for the company, given the breadth of experience.

Julie-Ellen Acosta

Alina Alexeenko

Professor of Aeronautics and Astronautics
and Chemical Engineering, Associate Dean
for Undergraduate Education

About the Author

Alina Alexeenko is a professor of aeronautics and astronautics and professor of chemical engineering at Purdue University. She also currently serves as Associate Dean for Undergraduate Education at Purdue's College of Engineering. Her research expertise is in computational rarefied gas dynamics, kinetic simulation of gases and plasmas, coupled heat and mass transfer analysis in applications to high-altitude aerothermodynamics, microelectromechanical systems (MEMS) and micropropulsion, and vacuum technology.

Dr. Alexeenko received her PhD in Aerospace Engineering from the Pennsylvania State University in 2003 and was a Women in Science and Engineering (WiSE) post-doctoral fellow at the University of Southern California from 2004 to 2006. She joined Purdue University faculty as an assistant professor in 2006.

Dr. Alexeenko is an associate fellow of the American Institute of Aeronautics and Astronautics (AIAA) and served as Chair of the AIAA Thermophysics Technical Committee for 2016-2018. She has authored over 200 journal and conference papers and is a co-inventor on 5 patents.

Dr. Alexeenko has been collaborating broadly with industry firms on the design and improvement of lyophilization equipment and processes since 2008 and is a founding co-director of Advanced Lyophilization Technology Hub—LyoHUB, established in 2014.

Questions and Answers

Personal Career Insight

What inspired you to choose a career in Aerospace and what has given you the most satisfaction during your career?

I was born on the day of the first international docking in space. This was something that I heard many times about from my dad. I literally grew up under a poster of the Apollo-Soyuz program and space travel was something that has always fascinated me. The biggest satisfaction in my career comes from hearing from our students about amazing things they go on doing and seeing their eyes light up.

Work/Life Balance

Did the pursuit of a career impact your decision on whether or when to have a family? What was the impact of that decision?

I had my first child while still in college and I think that was the best thing that happened to me. I did not graduate in 4 years but the break led me to rethink what really interests me professionally. Taking care of a small child really helped put things in perspective, added joy and purpose.

Mentorship/Sponsorship

What activities have you engaged in that have helped other women achieve success in their aerospace careers?

I helped our Women in Aerospace student organization get established a few years ago and serve as a faculty advisor. I really enjoy helping our very talented student leaders organize the Amelia Earhart Aerospace Summit at Purdue (now planning the third in March 2021). It's a celebration of Amelia's legacy in opening doors for women in aviation but is also aimed at empowering ALL for today's careers in aerospace.

Avoiding a Stall

How did you develop organizational savvy? And how did it help your career?

I am not sure if it's savvy (the shrewdness and practical knowledge), but I think it's always good to be curious of the Who and How of your organization. Every human organization has its own history and sometimes things don't make sense unless you learn how they got that way and not another.

With respect to your career, did you ever hit an organizational roadblock? How did you overcome it?

I believe in the theory of Beautiful Constraint. A roadblock or difficulty is an opportunity to transform something, sometimes your own point of view. In other cases, it illuminates other people's needs you have not understood.

Have you ever taken a role you were not excited about but had to show you were a "team player"? What was the outcome?

It happened at least a couple of times. Once very early in my tenure-track at Purdue, I was asked to take an extra non-research role in a big research center. I was not super-excited about it at first, but it turned out to be an opportunity to work closely with the center director who is an amazing mentor and to get to know more researchers from other groups and better understand their work.

Powering On

As you hit career obstacles, what motivated you to keep going?

(1) Kids, (2) Kindness of people who have removed obstacles on my path before I could even see them, (3) "Keep your thoughts raised high…" and the rest of Cavafy's poem "Ithaka." I keep a collection of songs, poems, and Bible verses handy to help get over a difficult day or a stormy season at work.

Kim Ashmun

Director, Global Sustainment
Lockheed Martin

About the Author

Kim Ashmun is the Director of Global Sustainment for Training and Logistics Solutions (TLS). She leads the TLS Global Sustainment Team responsible for developing strategy and driving execution to create strategic growth for the line of business. Prior to her current assignment, Kim worked for Lockheed Martin Aeronautics Company as the Director for the F-35 International Supply Chain. In this role she was the primary F-35 interface to the international supply base including BAE Systems Teammate and Alternate Mission Equipment. Kim and her team were integral to the development, implementation, and overall execution of the F-35 international supply chain management strategy and meeting the program affordability objectives.

Kim Ashmun

Prior to this assignment, Kim was the Director for the Air Vehicle, Missions Systems, and Vehicle Systems Supply Chain Management where she was responsible for leading subcontracts across multiple program life cycles with critical financial, schedule, and performance ramifications to Lockheed Martin Aeronautics and the F-35 Program. Ms. Ashmun also managed the F-35 Supply Chain Operations, where she supported F-35 business pursuits and led supplier proposal development efforts.

Ms. Ashmun also served as the Executive Project Assistant to the Chief Operating Officer of Lockheed Martin Aeronautics. She was responsible for leading special assignments and overseeing strategic initiatives for the enterprise. She assisted in critical integration activities across the enterprise, ensuring objectives were accomplished.

Previously, she served as the Production Non-recurring, International Participation Integration Lead, where she established technical planning and assistance expectations internally and across the supply base in alignment with the F-35 Best Value criteria.

Ms. Ashmun began her career in Engineering where she worked as a Manufacturing Engineer in Model Operations and Test prior to being selected for the company's Engineering Leadership Development Program (ELDP). During ELDP she worked in F-35 Air System, F-35 Basing and Ship Suitability, and F-35 Vehicle Systems Operations. Upon completing the ELDP program, Ms. Ashmun joined the F-35 program as a Systems Engineer on the F-35 Vehicle Systems Operations Team.

Ms. Ashmun has a Master of Business Administration and a Master of Science degree in Systems Engineering from Southern Methodist University in Dallas, Texas. She has a Bachelor of Science degree in Engineering Technology, Manufacturing

Engineering from Texas A&M University in College Station, Texas. She joined Lockheed Martin in May of 2002.

Questions and Answers

Personal Career Insight

What was the coolest thing you experienced in your career?

When I graduated from college at Texas A&M University, the contract for the F-35 System Development and Demonstration program had recently been awarded to Lockheed Martin (LM), and I had accepted an offer from LM. I was very excited about the job itself and the opportunity to work on this once-in-a-lifetime program. Shortly after embarking on my career with LM, I was selected to participate in their Engineering Leadership Development Program, which gave me the ability to work in multiple areas within engineering over three 1-year rotations.

There was no shortage of amazing projects, from supporting low-speed tests on the F-35 STOVL (F-35B) model at DNW (German-Dutch Wind Tunnels) in the Netherlands to nozzle calibration efforts at NASA Langley Research Center, but, by far, the coolest thing I got to do was during one of my rotations on the Basing and Ship Suitability team. I will never forget boarding the USS Harry S. Truman, a Nimitz-class aircraft carrier with a team of engineers to evaluate ship capability for the carrier variant of the F-35 (F-35C). It was just me and one other female on the engineering team, and she was more experienced in the role.

At the start of the trip, we paired up and she took me under her wing. She provided guidance and coaching, enabling me to find my own path and add value to the team. She offered to be a mentor to me and one of the first females I remember reaching out and genuinely wanting to help me in my career. She was a great mentor and I learned a lot from her technical expertise. It was her guidance and willingness to share her experience with her own career path with me, however, that really motivated me to rethink my own career goals early on.

The trip was a great experience for me early in my career and has shaped who I am as a leader today, and how cool was it that I got to catapult off the aircraft carrier!

Work/Life Balance

Did the pursuit of a career impact your decision whether or when to have a family? What was the impact of that decision?

I have always been a very driven person, whether it was in school, sports, or work. When I started in the workforce after graduation, I was determined to go places, and nothing was going to get in my way. My career was very important to me, and my husband, Travis, was very supportive of that. I put work first over everything else, always. The work that we do at Lockheed Martin is innovative and challenging, which motivated me to work hard and immerse myself in multiple projects concurrently.

At the same time, I kept telling myself I would know when I was ready to have a family. It was even a possibility that kids weren't in the picture for me and my husband. I thought that to be successful at work, I had to focus only on that and sacrifice having a work/life balance. I eventually realized that although I really enjoyed what I was doing and my job was very important to me, it's the experiences in life that really define us, and I didn't want to be defined only by my career. There was no reason I couldn't have both a family and a successful career. I was artificially limiting myself based on what I perceived I needed to do as a female wanting to excel in the workplace.

There is a balance, which is why it is called work/life balance, and the tipping point is at a different place for each individual. I think when people come to the same realization at some point in their life and career, each can decide what is best for them and what the right timing is. But I would encourage everyone to think about it sooner, I wish I would have. Travis and I have two wonderful boys, Kai (5) and Rhys (2), and I can't imagine not having them in our lives.

Looking back, I think career planning would have also been important for me versus what sometimes felt like relying on "everything just seems to work out." When I am asked for advice from early-career professionals, career planning is at the top on my list. It is so important to help provide direction and help someone navigate through their career and, most importantly, make informed decisions about what's next and how that next step can help you achieve your professional and personal goals.

Mentorship/Sponsorship

How important was mentorship/sponsorship for your career? Have they been men or women? How was the relationship established?

Throughout my career, I have learned the importance of sponsorship. I have had both men and women that were great leaders, and some not so great leaders. I realized that it is important to be the leader you want to be and understand what that means for you. And that good and bad leaders can help us shape what "great" looks like for us. I have also learned that there is a big difference in having a leader that is a supporter versus an advocate. I have had good leaders that supported me but really weren't an advocate and, ultimately, held me back from progressing in my career.

This is a hard lesson, hard when you learn it and hard sometimes to identify which type of leader you have. Find those leaders who are truly advocates for you because those are the leaders who will be honest with you, won't hold you back, and only want what is best for you and your career. My leader today does just that. As a leader, I try to be that advocate for the people who I believe in.

One time I had a leader who I was very excited to work for, and I will never forget my interview with him. We talked over the allotted interview time and it was one of the easiest conversations. He took a genuine interest in what I wanted out of the role and how he could help. I told him I knew I could do a good job. He quickly stopped me and said I was here because he knew I would do a great job, and it was not something I needed to prove to him. My experience working for him was one of the best in my career. He was an advocate for me and he taught me what it means to be a strong but fair leader and rise above, never losing clarity and focus on what really matters to you in your career.

Avoiding a Stall

How important is an advanced degree?

I am an advocate of continuing education; I have completed two advanced degrees. However, the importance of an advanced degree and the value that it brings to your career really depends on what you are doing and where you want to go. I came into Lockheed Martin with a Bachelor of Science degree in Engineering and went into an engineering career path. As I progressed, I realized that to advance on the technical path in my career and really get a depth of experience in engineering, I needed to get a master's degree. I ended up getting my Master of Science in Systems Engineering while continuing to work full time at Lockheed Martin.

Although sometimes challenging, I really valued being able to apply real work experience while getting the degree in parallel. I am often asked by mentees how do you make the decision to stay technical or go into leadership. In my experience, building that technical depth was important to me, but I also enjoyed the business side of things. I really spent the time to understand what energized me at the end of the day and it was solving complex business problems with the team. And that is why I decided to go back to school and get a second masters.

This time I focused on what degree would be best to support where I wanted to go next in my career. I wanted to go into program management, and I understood the importance of understanding the fundamentals of the business to be a successful program manager, so I pursued an MBA. This was probably one of the best decisions I made in my career. Not only did I enjoy the work, but I was getting the degree alongside my colleagues, I enjoyed teaming and challenging one another and, in the end, built a great network.

Ultimately, as I said, it is crucial to get a degree that parallels your career trajectory. I see time and time again people in business roles pursuing advanced engineering degrees because they think any advanced degree will help their career at an engineering company. It may provide a quick spike, but it is lost in the long term if the degree has little to do with performing your job.

Powering On

Did you ever feel limited in a role and how did you circumvent that?

I spent about eight years in engineering taking on different roles with increasing levels of responsibility and I enjoyed what I was doing. I found my work very rewarding, but I knew that I wanted to get into leadership and eventually into program management. To do that, I would also need to get experience outside of engineering, which was a big decision for me.

Making this career change would force me to go do something outside of my comfort zone. I had done relatively well and built a great network in engineering, and when I told my leader I wanted to do something different, outside of engineering, I didn't get the initial reaction or support I was expecting. This was another lesson learned for me, no one is going to manage your career for you, **you** have to manage it. Sometimes leaders look out for their own best interest, and leaders want to keep good performers. Great leaders can look past this and advocate for what is in the best interest of the employee.

When I made the decision to change roles, I did it not knowing what was in store for me next. I eventually ended up pursuing an opportunity in supply chain management (SCM), something I knew very little about. It was the network I built in engineering, the technical expertise I brought from engineering on process and procedures, and the skillsets I acquired over the years that helped me be successful.

I ultimately spent eight years in supply chain management when I anticipated only spending one or two. I gravitated to the business side, contracts, terms and conditions, strategy, and negotiations, all the things that I was motivated by. I became what I could have never envisioned, the subject matter expert and leader in SCM. I built a great organization and team and fostered great leaders within the company, some of the most rewarding experiences for me in my career.

Carina Greaves Birgersdotter

Principal Engineer
SAAB

About the Author

Carina Greaves Birgersdotter graduated from Linköping University, Sweden in 1983 with a Bachelor's degree in Education for teaching. After graduation, she moved to Italy where she lived for five years and learned to speak Italian. In 1988, she returned to Sweden to teach and to pursue a Bachelor's degree in Philosophy, Sociology, and Pedagogy at Linköping University. She graduated in 1990.

Carina started a family in 1990 and had two children by 1991. She continued her career as a teacher but knew that she wanted to do something different with her life. She chose to realize her dream of becoming an engineer, and in 1997, she spent the year preparing to begin her engineering studies. In 1998, she entered a Master's program at Linköping University where she studied Science in Media Technology and Engineering. She graduated with her Master's degree in 2004 and started working at Saab AB in Linköping. This site manufactured aircraft, not cars, and she worked on military aircraft.

Carina Greaves Birgersdotter

She started her career at Saab AB, as a Systems Engineer on the Aircraft Central Health Monitoring System program. She was promoted to Team Leader, to Head of the Aircraft Central Health Monitoring system, and then to Principal Engineer. She worked with Central Health Monitoring, Registration, Testing, and Function Monitoring; her main interests leaned towards Functional Monitoring. She was passionate about ensuring the data collected and presented to the pilot was clear and easy to understand so that the pilot knew what action is needed to be taken. She understood the importance of ensuring that critical information was not lost in a plethora of data generated by secondary failures. She worked directly with Saab test pilots and also pilots and maintenance staff located at military flight bases to achieve these important goals. She realized that these goals were extremely important in order to optimize the Aircraft Health Management System on these aircraft. In 2017, she left Saab AB and Sweden to move to the island of Madeira.

Carina became an active member of SAE International in 2017 and joined the HM-1 Committee for Integrated Vehicle Health Management. She served as the working group lead for "AIR6915 Human Factor Considerations in the Implementation of IVHM," which was published in 2020.

Questions and Answers

Personal Career Insight

How did you decide between a leadership vs. technical career track?

Before I studied to be an engineer, I was a teacher, which in itself is a leadership role—to lead students to knowledge. When I changed my career, the desire to be able to delve into technology was greater than teaching. I wanted to figure out technical solutions. I had no desire at all to lead others at work. I simply wanted to work technically myself. Women often become project leaders and leaders in other areas where they have more social leadership because men seem to think that women are good at it, taking care of and looking after others. I invested in becoming a technical leader, which I became in my field. I have always been interested in how things work, and one of my favorite subjects at school, when I was a child, was mathematics.

My daughter was the same when she was growing up. She is also now an engineer. I strongly believe that we women have an important role to play in supporting our daughters so that they can choose what they are interested in. When I was young, things were different. When I wanted to make my career choice and I said I wanted to go for a technical education, the person who would advise me before further studies just laughed at me and said that women should become medical nurses or teachers. I could not stand to see blood, so there was only one choice left. But I got my revenge later. Unfortunately, I missed several years in the profession, but hope that I made up for it.

Work/Life Balance

Did you ever feel guilty for leaving work on time to attend a family event?

I worked in Sweden when the children were younger. There is no major difference between men and women when it comes to taking care of children and family. I've never felt guilty. It happened just as much that the men needed to leave work earlier due to family matters. So there was an understanding of this. If you are a parent of small children, both domains, work and free time, must work together, which, I think, we are well on our way to in Sweden. It was almost as if it was more acceptable to go home earlier for the sake of their children than to go away earlier for a dental visit. Likewise, there was no major difference in who was at home when the children were ill. It was arranged in the family who stayed at home depending on the workload that day.

Every parent in Sweden has the right to take 240 days off work to care for each child until that child reaches the age of 12, for a total of 480 days per child. Men also have 10 days off in connection with childbirth. I have not met any man in my profession in the last 20 years who does not take his leave. The fact that the man has the right to be at home without the employer being able to protest also creates an understanding of taking care of the family and children that we women benefit from. It has also been seen in Sweden that men become better at working, collaborating, etc. when they get the opportunity to get more involved in the family.

Mentorship/Sponsorship

How important was mentorship/sponsorship for your career? Have they been men or women? How was the relationship established?

When I started my work, I had a female mentor. She was very nice and helpful, but she did not work in the field I was going to work in. So it became more and more that the mentorship went over to a man, who was also my boss. I think I was lucky. This was a man who encouraged and supported me. He always made sure I did the job first and made contact with the right people, but he stayed in the background. He empowered me and gave me the responsibility to do the job. When I felt ready, we discussed the results. Sometimes we came to the conclusion that I was thinking wrongly, sometimes I was thinking correctly.

I also felt that I could always ask if I was unsure and he helped me get onto the right track. There was no thought in his mind that my question could be stupid or strange. His door was always open. He encouraged me to learn more and "climb" in my career. The most important thing about mentoring is to have a mentor who believes in you and is knowledgeable about the field in which we are working, regardless of whether I am a man or a woman. This also applies regardless of whether the mentor is a man or a woman. In fact, in my career, I have had better male mentors than female mentors. In my view, the sex of the person does not, or should not, determine capability.

Avoiding a Stall

How important is an advanced degree?

For we women, higher education is very important. As the situation appears today, we must always show that we are better than men at performing the same task. Higher education gives us confidence and, in certain situations, can be an important differentiator. Engineering professions are dominated by men worldwide. There is inherently a tendency when considering an equivalent man or woman for career advancement that the man is selected. Men have a strong tendency to work together and support each other, whereas we women tend to work alone. We do not sufficiently support each other. I think this comes from our Stone-Age brain. Men hunted together, and if you wanted to be successful in the hunt, you had to work together. Women were at home guarding the children and their home. No other woman could enter there and possibly take the man away because in this case you would be left without protection.

I have experienced in my profession that women are more oriented towards protecting their own field. It's sad that it's so, but it's probably something we women need to think about and work with. This would take time, as it is not easy to change human nature. Education is critical to this, but we must remember that this starts in the home, and we should all encourage our daughters that all opportunities are open to them.

Powering On

Was there a significant event that changed your career trajectory and what was it?

I have advanced four times during my career. From Systems Engineer, Team Leader, Head of the Aircraft Central Health Monitoring System, and finally Principal Engineer. I enjoyed my positions, but then I got a new manager who was a pure administrator, with no technical knowledge, and who thought I should try something else since I had been in the same area for a long time. She thought that I should leave this domain so that someone else could make a career. And this person was a man who had complained that he had not been able to move on in his career.

She said that I should leave my position and at the same time teach him how to do the job. I felt that I would be left behind by a man with less knowledge than me. I was also skeptical of the explanation. I would have accepted him to work with me on equal terms, but in this way, he would have a higher position than me with less knowledge. I was not interested so I started looking for other work. I submitted my resignation, but my boss's boss asked me to take a sabbatical for half a year, and he hoped I would come back after a couple of months. But after the sabbatical, I quit, and the man who wanted my job got my job. After about six months, he was relieved of his duties because he could not do the job. I was not interested in returning after such treatment.

I have been able to maintain my interest in my domain through continuing work with an SAE International technical committee.

Lisa B. Callahan

Vice President and General Manager
Commercial Civil Space
Lockheed Martin Space

About the Author

L isa B. Callahan is Vice President and General Manager of Commercial Civil Space at Lockheed Martin Space and has one of the coolest jobs in and out of this world! Whether it's returning humans to the moon, gathering samples from asteroids, exploring the planets of our solar system, or peering back in time with the Hubble telescope, Lisa leads the teams who are building and operating the spacecraft that are unlocking the riddles of our universe. In her role, she oversees the development of the next human lunar lander and development of the Orion spacecraft, part of the Artemis program, which will return humans to the Moon and eventually take them to Mars and beyond. Her team develops and operates the robotic spacecraft that are exploring and unlocking the secrets of the solar system and our planet, Earth. Lisa also oversees the development of the GOES weather satellites, which are significantly improving the way that forecasters predict severe weather and fires across the western hemisphere. Chances are you or someone you know has benefited from the forecasts enabled by these satellites. Lastly, her team connects people across the globe with satellite communication solutions.

Lisa B. Callahan

Previously, Lisa has held additional leadership positions at Lockheed Martin including Vice President of Corporate Internal Audit, Vice President and General Manager of the Mission Systems and Training Undersea Systems line of business, Vice President of Maritime Ballistic Missile Defense Program, and Program Director for Simulation Training and Support business.

Lisa graduated from Virginia Polytechnic Institute and State University with a Bachelor of Science in Electrical Engineering. Currently, she is a member of the Engineering Advisory Boards at the University of Southern California and the University of Colorado—Boulder. Lisa is married to Bob Callahan and they have two children, Carly and Ryan. In their free time, they foster puppies with a local rescue and have found homes for more than 500 dogs.

Questions and Answers

Personal Career Insights

What was the coolest thing you experienced in your career?

When I was a college student, I had no idea that a degree in engineering could lead to so many cool, memorable opportunities, even as a young professional. But I am proof that life as an engineer is much more exciting than I ever believed.

Very soon after starting my career, I had the opportunity to ride on a fast attack submarine—this included working, eating, and sleeping on board. In fact, I slept on a torpedo rack next to the weapons launch systems—something I won't ever forget because I'm not sure if they put me next to the weapons system, which was guarded

24/7, to keep me out of trouble, or because I was one of the first women the U.S. Navy allowed on a submarine and was, in fact, the only woman on board. That was an experience I will never forget.

While it was a remarkable opportunity, the trip wasn't only for fun, it was intended for our team to see how our customer was going to use the combat systems and for the U.S. Navy to test its initial capabilities. What I learned was that some things we believed to be important to the customer were not, and how they functioned on a submarine impacted how they used our systems in ways we had not previously understood.

Later in my career, I was building live training systems for the U.S. Army at the National Training Center in Fort Irwin, California, a training range in the desert about the size of Rhode Island. Our team's job was to design new systems for training soldiers, and we had a lot of great ideas that leveraged new technologies. To best understand how the trainers would use our systems, I spent a day and a half driving around a desert in a Humvee with no doors and no windshield. It was very hot and dusty during the day and extremely cold at night.

As I watched how the trainers use their current systems, I quickly learned that some of the concepts we wanted to implement were good ideas but were not going to work in the desert. For example, the reason the Humvees didn't have a windshield was because the desert sand builds up on the windshield and you can't see—so it was better to remove them, which meant anything we gave the trainers to use needed to be able to withstand the sand. Similarly, our team was looking at how we could give the trainers notepads with stylus pens to collect data and document findings as opposed to writing down information with pen and paper. What I found was that because it gets really, REALLY, cold in the desert at night the trainers wore big gloves, which prevented them from both maneuvering a stylus and using a touch screen on a tablet.

Both the submarine and desert experiences impacted me professionally. Not only were they fun and unique, but they reinforced the fact that as a professional, I had to do more than simply know our customers' requirements, I had to find a way to put myself and our team in the shoes of our customers to truly understand how to provide them the best solutions for their needs.

I've had other cool experiences as well, including participating in missile defense test flights from Hawaii and witnessing spacecraft launches from Kennedy Space Center. My engineering degree and my jobs have allowed me to literally travel the world and meet some incredible people.

Work/Life Balance

Did you ever have to make a move to advance your career (within your company or changing companies) that impacted your family life, and how did you balance the two?

Advancing my career at Lockheed Martin has meant moving my family around the country several times. Some of these moves were easier than others, and the challenges increased as my children got older. In 2008, we moved from Florida to New Jersey, which was a good move for my family. When I received the job offer, I had time to sit down and discuss the opportunity with my husband and kids. They were all excited about the change and my husband was able to keep his job and work virtually. This move turned out to be great for all of us, and it helped that we were able to plan for it in advance and figure out how to balance the needs of our kids and our careers. That was a big deal.

Unfortunately, the next time I needed to move, it was in the middle of the school year and we didn't have time to plan for the change. My family, especially my children, were very happy living in New Jersey and not really interested in moving to Virginia. But we made the decision to go, and we had to make the best of it. One way I helped my children adjust to the change was finding soccer teams for them to join near our new home. Being on a team helped them connect with other kids who share their interest in soccer, which also made it easier for them to transition into a new school environment when classes and friend groups were already established. However, the curriculum in the school system differed and both children struggled academically. Moving in the middle of the school year was hard on everyone.

When we moved to Colorado the biggest challenge was acclimating my son. My husband was continuing to work virtually, and my daughter was leaving for college. My son was already involved in a lot of online gaming with his friends, so when we moved to Colorado, he maintained his friendships through this outlet. At the time, I thought it was a good thing. In hindsight, I realize it did not help him to create friendships in Colorado.

While moving my family always had challenges, I truly appreciate that with every new job, new energy is sparked in me. I enjoy learning new things and building new relationships. New jobs force me to learn new skills and figure out how to navigate in new cultures and environments. Today I benefit from all these experiences and can draw upon them to find solutions to challenges I may face or to anticipate the future. Relocating isn't good for every family, but as it turns out, these changes were good for my family, although they didn't always feel that way at the time. My children learned they can adjust to change, to new environments, and to making new friends. As a result, they are comfortable taking on new challenges. As a mom, it is great to see how these changes have shaped who they have become as adults.

Mentorship/Sponsorship

How important was mentorship/sponsorship for your career? Have they been men or women? How was the relationship established?

I believe mentorship is very important for everyone, and both mentors and sponsors played a critical role in helping me throughout my career, but in very different ways.

A mentor is someone who can talk through the challenges you're having, can give honest feedback on your ideas and pursuits, and can help you navigate your career. I have always had many mentors; in fact, I often encourage people to have a personal board of directors (BoD) of mentors. When I think about who should be on my board of directors, I think about where I have holes in my personal knowledge or network. For example, if I am starting a new role, I will look for someone who understands the organization I have joined. I may also look for someone who understands the technical solution we are developing and someone who understands our stakeholders. These are most likely different individuals who I reach out to and ask them to be mentors. I encourage others to use this BoD concept even if you are only looking for career advice because getting diverse perspectives is critical. We all take different paths to reach our career goals and there is no set way to achieve your desired outcome, so these diverse opinions are important.

Sometimes mentors are previous bosses, sometimes they are peers in your organization, and they can even be someone from a younger generation in the workforce. For example, I recently I had a mentor who was a young professional who helped me navigate the latest collaborative technologies that were important to our workforce as we moved to a more virtual workspace due to the COVID-19 pandemic. Many early-career professionals grew up using online platforms such as Slack and Zoom, I didn't, and yet I needed to know how to use these programs to effectively to lead our team and assure collaboration. My mentor was able to teach me these technologies and help me incorporate them into my everyday life.

Sponsors are completely different from mentors. While you ask someone to be your mentor, you earn sponsors through your performance, and ultimately, they end up advocating for you when you are not in the room. At Lockheed Martin, I witness sponsorship often during our talent review meetings. We hold these meetings to discuss talent development and succession planning. Usually, the people we discuss are not in these meetings, but they have advocates in the room who know of an individual's experience, leadership and technical skills, and career aspirations. These sponsors are critical in helping individuals advance their careers, and I know this has been true for my career. There are certainly previous bosses that I know have sponsored for me, but throughout my career, I have learned I had sponsors I didn't know were sponsors. For example, when I interviewed for the job in New Jersey, which led me to transition from Florida to New Jersey, my selection was accomplished due to the strong sponsorship I had advocating on my behalf. The leadership team in New Jersey didn't know me; however, my sponsors convinced them to take a risk by bringing me into the organization.

I said you earn sponsorship, which I believe is true, but there is no reason you can't ask for sponsorship when you are competing for a job. Think about who is the hiring manager, is there someone you know who influences that hiring manager? Could you ask that person to put in a good word for you? What do you have to lose? If they say no, you can ask for some valuable feedback that you can address for the next role.

Avoiding a Stall

Was there a significant event that changed your career trajectory and what was it?

If I were to point to a single event that significantly impacted the trajectory of my career, it would be my move from Florida to New Jersey. I had been working in Orlando with Lockheed Martin's training and simulation organization for 15 years. I started as a new engineering manager in the organization and grew my career over those years to become a Program Director for our constructive and live training business, managing a portfolio of programs. I had built strong relationships within the Lockheed Martin organization in Orlando and with our training and simulation customers.

After 15 years, I was asked if I would like to interview for a leadership role in Moorestown, New Jersey, leading the Aegis Ballistic Missile Defense programs. I felt this was the biggest risk I would be taking in my career. I didn't know anyone in Lockheed Martin based in Moorestown, and I realized much of my success in Orlando was due to strong professional relationships and knowing who to ask for help when I ran into a challenge. I also didn't know anything about the Missile Defense program, or the technologies used to support the program. After talking this decision over with my mentors and my family, as it obviously required us to relocate, I decided to compete for the role and was ultimately offered the job.

I was successful with this career move for many reasons. To start with, I had a great leader. He sat me down and helped me understand the status of the business, shared that he wanted me to spend my initial months on the job learning from the team and that my program management experience was a strength I could offer the team. I was also extremely fortunate to have a great set of peers who welcomed me and helped me get to know the organization and the customers. Lastly, the team I was leading was outstanding and really had great customer relationships and strong technical skills. They too welcomed me and spent valuable time educating me as I learned the business. Ultimately, I was able to successfully integrate into this business and build relationships within our organization and with our customer community. Many of these relationships I still maintain.

The reason I believe this was a significant trajectory in my career is twofold: First, it built my confidence in my ability to move into a new role, with new people and new technologies, and to adapt, learn new skills, and impact the business. Second it demonstrated to the Lockheed Martin leadership, my ability to make changes and lead effectively in new areas. This benefited my career as it led to additional opportunities to change roles and broaden my experience base and continue growing my career.

Powering On

Have you ever taken a role you were not excited about but had to show you were a "team player"? What was the outcome?

I can think of two roles in my career where I was asked to take on a role that didn't initially excite me. The first involved me leaving my program director role in Orlando to lead a change initiative, and the second was a corporate assignment. While I was reluctant in both cases, I was willing to take the roles because they were important to the organization, and my mentors helped me see how I would also benefit from each job. In both cases, I believe the jobs ended up being good opportunities for me for several reasons. First, they demonstrated I was a team player and would support what the organization needed of me. Second, I learned a lot from each of these roles, more than I initially expected. The corporate role taught me about parts of our company I had not been exposed to previously and greatly expanded my network. I learned a lot through the change initiative role as well, which has greatly improved my skills as a leader and my ability to set visions and effectively communicate and align organizations to achieve change. Furthermore, by taking on these roles, I learned more about myself, discovering the type of jobs that excite me and helped me to determine my future career aspirations and goals.

There are two important lessons I would share with others facing these types of assignments. The first is to find ways to incorporate your passions and strengths into the work you are being assigned. When I went to the corporate assignment, after I was feeling comfortable with my new responsibilities and felt I had the bandwidth to add projects, I asked my boss to include me in a special assignment that allowed me to leverage my skills in strategic thinking. This extra assignment helped use my passion for strategy in a new and exciting way while driving positive outcomes for the corporation.

I also learned it is important to be transparent and communicative with your leader regarding the expectations of these roles. I made sure to ask my leaders what they expected of me in these roles and the impact on the business I was expected to contribute, while also sharing with them what I hoped to get out of the role. We also discussed the expectations of how long my tour of duty might last. This type of communication was extremely beneficial and helped ensure that there were common expectations, effectively avoiding any misunderstandings or disappointments.

Mary Anne Cannon

Vice President
West Palm Beach Florida Site and
Development Operations
Pratt & Whitney

About the Author

M ary Anne Cannon is Pratt & Whitney's Vice President of the West Palm Beach Florida Site and Development Operations. Mary Anne is responsible for oversight and leadership of the West Palm Beach campus, inclusive of anticipated growth, and leads the Development Operations organization across all Pratt & Whitney sites, encompassing Engine Assembly, Instrumentation, and Test Operations. She is responsible for meeting Engine Assembly, Instrumentation, and Test-related customer requirements in the areas of cost, quality, schedule, and delivery, as well as developing organizational talent. Prior to this role Mary Anne was Vice President of Operational Commercial Programs and was responsible for developing and managing the product line life-cycle strategies and ensuring customer requirements were met for multiple engine product lines. Ms. Cannon has 30 years of experience at Pratt & Whitney across Engineering, Quality, Operations, and Program organizations.

Mary Anne Cannon

She served on Pratt & Whitney's executive committee as Vice President of Quality and Environment, Health and Safety (EH&S). She was responsible for the continued development and administration of Pratt & Whitney's Quality and EH&S policies and practices worldwide, as well as ensuring compliance with corporate, federal, and state requirements, product integrity and quality assurance, and Achieving Competitive Excellence (ACE) continuous improvement activities. Prior to joining Pratt & Whitney in 1989 at the West Palm Beach location, Cannon started her career at Grumman Aerospace Corp. She earned a Bachelor's degree in Mechanical Engineering from Stevens Institute of Technology in 1986.

Mary Anne is passionate about supporting women in STEM (Science, Technology, Engineering, and Math) roles. She devotes her time to mentoring young women both in college and at Pratt & Whitney, and most recently she and her husband established a scholarship at her alma mater Stevens Institute of Technology to support women pursuing an engineering degree. Mary Anne serves on the Presidents' Leadership Council for Stevens Institute of Technology. She has served on the Industry Advisory Board for Western New England University School of Engineering and the foundation board of Central Connecticut State University.

Mary Anne is a lifetime member of AWAM (Association of Women in Aviation Maintenance) and a 20-year member of Women in Aviation International.

When not at work she can be found at the gym pursuing her newest hobby—powerlifting—most recently she competed in her first amateur competition and placed first in her category.

Mary Anne makes her home in Stuart, Florida, with her husband Dan—who recently retired from Pratt & Whitney after a 35-year career and her son Joseph who recently graduated with an AS in Criminal Justice and is pursuing his BS in Criminal Justice.

Questions and Answers

Personal Career Insight

What inspired you to choose a career in Aerospace and what has given you the most satisfaction during your career?

I'm not sure that I chose a career in Aerospace but, rather, found my way to this amazing and exciting field through a series of events. My parents encouraged my brother, sisters, and me to be curious and explore the world around us. I grew up in a small beach town in New Jersey, and we spent our summer days like most kids in the 1970s and 1980s—but we loved to solve problems, often in the form of word games, crossword puzzles, and cryptograms.

We loved to help our father fix things around the house and work on cars. It was FUN. I loved going to the library and reading about new places and scientific discoveries. Growing up, Jacques Cousteau was one of my heroes, as was Dr. Robert Ballard (long before the Titanic discovery). Every year we looked forward to a family favorite: an Easter egg hunt where my father would spend the months leading up to it writing riddles for each of us to solve that would guide us through the egg hunt.

So when it came time for college, there really was only one school and one choice: study Mechanical Engineering at Stevens Institute of Technology in Hoboken, NJ—where my dad received his master's degree in the early 1960s. I truly thought I would spend my career working in a power plant, but as fate would have it, 1986 was not a great job market year. I graduated with a Bachelor's of Engineering degree, majoring in mechanical engineering, and sent over 100 resumes and job applications with no luck.

A very close friend of my father who worked at Grumman Aerospace invited me to visit him at work in Calverton, NY. He arranged a few job interviews and I received an offer to be a flight test instrumentation engineer. That summer I had fallen in love with the high-speed flying of the F-14 in the movie Top Gun. Driving onto the Grumman plant site was like driving onto a movie set; from my first day on the job, I was bit by the aviation bug.

Early on in my career, I came to work for Pratt & Whitney, and 34 years later, I love being part of the aerospace industry and have built an amazing career. Being part of the team that developed the F119 engine that powers the F-22 has and continues to be one of my most satisfying career moments.

These days my most satisfying moments are spent mentoring and coaching and inspiring the next generation. I love to tell the next generation, "Don't limit yourself; none of us know the path your career is going to take, and if things don't go according to your plan, don't take it as a setback."

Work/Life Balance

Did the pursuit of a career impact your decision whether or when to have a family? What was the impact of that decision?

I spent the first decade of my career thinking I had to make a choice between having a career and having a child. Quite honestly, my son was not a planned pregnancy, but he's the best thing that's ever happened to me.

In 1998, my job required a great deal of travel, and I still remember when I had to tell my supervisor that I was pregnant and my doctor had advised me to stop traveling. I was very nervous about having to tell my supervisor of my pregnancy and limitations for travel, and I was very concerned that my career would be over. Both my partner and I worked for Pratt & Whitney at the time, and during my maternity leave my partner took a new position. The company treated me as "a trailing spouse" and found a position for me in San Antonio so I could be with him. So I entered my maternity leave in Florida, and when I returned to work, it was in San Antonio.

When I look back on this time in my life and career, I realize that the thought of having a baby would damage my career were self-imposed thoughts, Pratt & Whitney was great to me.

In the 20-plus years since then, the workplace has changed, but for some women, it may still be terrifying to give maternity notice.

I've had quite a few women discuss with me their thoughts before making a decision to start a family. I tell them the challenges are real when it comes to managing a family and career, but if you want to be a mother, then be a mother; you absolutely can still have a career.

Mentorship/Sponsorship

How important was mentorship/sponsorship for your career? Have they been men or women? How was the relationship established?

I've benefited from both mentorships and sponsorships during the course of my career, and in line with the workplace at the time, all of my mentors were men. Often mentors were people I sought out for advice and support in my career. My first mentor has since retired from the company, but we still connect occasionally. This relationship started when he was my supervisor, but over the 25 years, as I moved around the company, we continued the mentorship.

Sponsorships have all developed organically after establishing a work relationship based on trust and understanding. It's also important for sponsors to understand your career goals to help advance them. Sponsorship became very important when I wanted to transition into the executive levels.

In fact, the first executive job I got was due to the sponsorship of the head of our military engines division at that time, and it was then that I realized how critical it can be to a career. I was interviewing for an executive-level job in our commercial engines division and the interviewer asked "Why should I give you this job—you have no commercial experience or experience in aftermarket—but so-and-so said you would be great for this role." That's when I realized how critical a sponsorship was to advancing my career.

Now that I'm in the position to act as a mentor or sponsor to others, I take it very seriously. It can be employees who work on my team or employees in other divisions of the company. I do have a special focus in wanting to make sure women and girls know the career paths available to them in aerospace, even beyond engineering.

Aerospace employs people in operations, programs, finance, human relations, and communications, to name a few.

Avoiding a Stall

Have you ever taken a role you were not excited about but had to show you were a "team player"? What was the outcome?

I was in my first executive position for five years. At the end, I felt like I outgrew it, didn't get exposure to other positions and felt like my career was stagnating. I chose to have a conversation with another executive who ranked above me and said that I felt ready for another position. I also questioned why I hadn't been on interview slates for a position that was an advancement from the one I was in at the time. The other executive responded that I was doing a good job and he didn't know I wanted to do something else.

A few months later I was approached to take a job in another organization that I viewed as outside my career path. Initially, and briefly, it made me angry because I felt that I was being distracted from my goal. But, ultimately, I'm a team player so I jumped enthusiastically into the role. Then almost one year to the day, I was offered the Vice President job I always wanted in our Quality group.

I learned that taking the job I didn't want gave me the exposure to other leaders in the company. Those leaders saw what I was capable of and then gave me the role I wanted.

It's important to advocate for yourself. You may not get the role you really want and it may not be on your timeline, but if you trust in the process it does work out.

Powering On

As you hit career obstacles, what motivated you to keep going?

I've learned not to see them as obstacles but challenges that become opportunities. Generally speaking, women let these self-doubts creep in and are harder on themselves than they are on anyone else. Over 30 years, I've learned to power on through the self-doubt and trust that there's a reason that other people believe in me and that I can do the job.

There have been times in my career where my vice president role was combined to lead two organizations, or I was appointed to a role that I didn't feel qualified for—especially when the president of Pratt & Whitney at the time had full confidence in me to take a role. Whenever I've had doubts that I couldn't do something, I've gone back to "If these leaders put me in a role and they believe I can do it, why don't I think I can?"

Have confidence in yourself and your abilities. It also goes back to building those mentorship and sponsorship relationships: if those people believe you can do something, then believe you can do it.

Charmaine Chin

Director, Global Cleanliness Supplier
Delta Air Lines, Inc.

About the Author

Charmaine Chin currently serves as Director of Supplier Strategy for Delta's Global Cleanliness Division. Her team is responsible for corporate-level partnership management and oversight for the strategic cleaning suppliers that support Delta aircraft, airports, and facilities across the world. This involves developing the contract strategy, building partnerships, reviewing supplier performance, and managing supplier transitions. Prior to this role, Charmaine held several positions of increasing responsibility within the company's Technical Operations division for the first 19 years of her career.

She started as an MD11 Fleet Engineer in 2001 and worked her way through technical ranks for various engineering teams supporting aircraft structures, interiors, engines, and components prior to transitioning into leadership in 2012. During her time as a leader in Technical Operations, Charmaine served as a B747/B777 Fleet Engineering Manager, General Manager for Atlanta Base Maintenance, and General Manager for Fleet Projects. Charmaine's time in aviation has provided experience in engineering, maintenance, program management, operational support, and vendor management. Her career journey helped to develop a unique combination skillset that positioned her for the current role in Supplier Strategy.

Charmaine has been an active member of the Cobb County Airport Advisory Board for five years, serving as Chair in 2018. She is the former Chair of the Delta TechOps Diversity Council and served as a member for four years prior to joining the Global Cleanliness team. She has a Bachelor of Science in Aerospace Engineering from the Georgia Institute of Technology and a Master of Business Administration from the Emory University Goizueta Business School (W18). She is also Six Sigma Green Belt certified. Charmaine enjoys spending time with family, coaching football, and hanging out with her four-legged fuzzy canine children (Samson and Luna) in her spare time.

Questions and Answers

Personal Career Insight

If you could give early-career aerospace women one piece of advice, what would it be?

Be bold, brave, and never stop learning. You are the person who is ultimately responsible for your career. It will be what you want it to be as long as you stay in constant pursuit of your goals and dreams.

Work/Life Balance

Did you ever have to make a move to advance your career (within your company or changing companies) that impacted your family life, and how did you balance the two?

There were several moves made to advance my career.

- Taking on roles to learn new skills and gain exposure to other business units within the company.
- Changing companies as a young engineer due to industry challenges
- Investment in higher education (Emory Executive MBA program) and external leadership development programs to gain higher business acumen and help promote growth

Career moves always have the potential to impact family life due to time constraints associated with new responsibilities. However, clear communication, staying committed to planned family time, and carving out dedicated "you" time are keys to maintaining a healthy work/ life balance. Useful practices include:

- Scheduling weekly date nights with your spouse
- Honoring set "be home by" times to maximize family time outside of work
- Limiting work related phone calls and e-mails during family time
- Giving advanced notice for any unforeseen change of plans
- Using active listening to talk through work-related misunderstandings
- Apologizing to your family when actions don't line up with intentions
- Setting daily personal time for mindfulness and meditation
- Planning your time (24/7/365)

Mentorship/Sponsorship

How important was mentorship/sponsorship for your career? Have they been men or women? How was the relationship established?

- Mentorship/sponsorship is a critical element for success. Career advancement to the senior levels of leadership is only possible through the support of the mentors and sponsors in your network. Performance alone is not enough. Mentors are advisors that stretch your thoughts through their wisdom and drive actions to help in your growth. Sponsors are people who advocate for you when you are not in the room and provide financial support for development opportunities. Both are important elements to give you the exposure to others that you may not have access to on your own.

- My mentors/sponsors have been both men and women. Men give additional perspectives that women may not have. In addition, their circles of influence may potentially expand further than the women at your company. Women mentors/sponsors are valuable resources that provide wisdom on building resilience, showing value in unique ways, & navigating male-dominated environments. They also have more access to women's leadership development programs and may be more nurturing than a male mentor/sponsor.

- The relationships were established common interactions during meetings, cross-functional projects, and structured company programs.

Avoiding a Stall

How did you develop organizational savvy? And how did it help your career?

Several ways:

- Switching to roles in other business units.
- Talking less, listening more.
- Being a part of cross-functional teams.
- Investing in my own personal and professional growth through internal and external programs (leadership development company programs, Emory Executive MBA, Executive Leadership Council Bright Futures Program, Pathbuilders Achieva Program).

Developing organizational savvy helped my career by:

- Providing knowledge on interdependencies within the company
- Giving clarity to areas of opportunity
- Providing exposure to decision makers and people of influence
- Improving strategic thinking skills
- Building confidence and expertise

Powering On

As you hit career obstacles, what motivated you to keep going?

- Faith
- Belief in myself and my ability
- Strong personal board of directors
- Having a competitive nature
- Controlling what I could control: my attitude, thoughts, actions, and behaviors

Nagin Cox

Tactical Mission Lead
Mars Curiosity Rover
NASA/Jet Propulsion Laboratory

About the Author

Nagin has been exploring since she decided as a teenager that she wanted to work at NASA's Jet Propulsion Laboratory (JPL). She was born in Bangalore, India, and grew up in Kansas City, Kansas, and Kuala Lumpur, Malaysia. Her experiences as a girl in a multicultural household showed her how easily we separate ourselves based on gender, race, or nationality, and inspired her to do something that brings people together instead of dividing them. The Space Program helps the world "look up" and remember that we are one world. Thus, she has known from the time she was 14 years old that she wanted to work on missions of robotic space exploration.

Nagin realized her childhood dream and has been a spacecraft operations engineer at NASA/JPL for over 20 years. Nagin has held leadership and system engineering positions on interplanetary robotic missions including the Galileo mission to Jupiter, the Mars Exploration Rovers, the Kepler exoplanet hunter, InSight, and the Mars Curiosity Rover. She was also involved in the MOXIE (Mars Oxygen ISRU Experiment) team, which will prototype making oxygen on Mars from the Martian atmosphere

Nagin's honors include being the namesake for Asteroid 14061. She has received the NASA Exceptional Service Medal and two NASA Exceptional Achievement Medals. She was an inaugural winner of the Bruce Murray Award for Exceptional Public Outreach. She is a U.S. Department of State's STEM Speaker and has spoken to audiences around the USA, Canada, Europe, South America, the Middle East, and South Asia on the stories of the people behind the missions. Her lecture on Mars Time on the TED Talk website has been viewed more than two million times (https://www.ted.com/talks/nagin_cox_what_time_is_it_on_mars).

She is a past member of Cornell University's President's Council of Cornell Women and has served on the boards of Griffith Observatory (FOTO) and Impact Personal Safety: Self-Defense and Empowerment for Women.

Before her time at JPL, she served for six years in the US Air Force including duty as a Space Operations Officer at NORAD/US Space Command. Nagin holds engineering degrees from Cornell University and the Air Force Institute of Technology as well as a psychology degree from Cornell. Sometimes she is not sure which one she uses more, the engineering degree or the psychology degree!

She is currently a Tactical Mission Lead on the Mars Curiosity Rover as well as the Deputy Team Chief for the Engineering Operations Team for the Mars 2020 Rover, and every day at NASA/JPL, exploring space is as rewarding as the first.

Questions and Answers

Personal Career Insight

What inspired you to choose a career in Aerospace and what has given you the most satisfaction during your career?

Sometimes your life path can be traced to a single idea. In my case, I was inspired to pursue a career in Aerospace by a realization that people frequently find ways to divide themselves, and it is much harder to find common ground. When I was growing up, I was confronted with the notion that different genders had different futures—that there were different expectations of me because I was a girl. Fortunately, I was also lucky enough to see firsthand part of the impact the Apollo lunar missions were having on the whole world. Even though I was young, it was still obvious to me that people around the world were gathering in front of their televisions to experience the missions together.

The wonder of what was happening was evident in large type in morning newspapers above the fold. That stood in stark contrast to the challenges I was experiencing in my culture with the assumptions of the role of women and girls. This dichotomy of humans voluntarily focusing on our differences instead of concepts that bring us together is evident throughout the world today. Robotic space exploration stands as a beacon of global cooperation, and my younger self knew this was the path for me. I have now been in the business of building and operating spacecraft throughout the solar system for multiple decades, and it is still that pursuit of exploration for all humankind that gives me lasting satisfaction.

Work/Life Balance

Did the pursuit of a career impact your decision whether or when to have a family? What was the impact of that decision?

My immediate family is a family of two—my spouse and I. After a rewarding stint in the United States Air Force and then as an engineer at IBM, I was thrilled to be offered my dream job at NASA's Jet Propulsion Laboratory in Pasadena, California. The offer came at a time when my family was living in Northern California. Thus, accepting the only job I ever wanted would involve a separation from my spouse. He was not surprised by this dilemma since, from our earliest days together, he was well aware of my single-minded aspiration to make it to JPL.

He began the process of relocating himself professionally to Southern California so we could be together. Given his skillset as an accomplished engineer, he has indeed forged a rewarding technical career based in Los Angeles. As with any family with independent careers, the years have been filled with discussions of goals, needs of our family, and an understanding that there might be periods of separations or commuting when we were based in different cities temporarily.

Mentorship/Sponsorship

What activities have you engaged in that have helped other women achieve success in their aerospace careers?

My future in engineering and space exploration would never have come to pass if it were not for my mother's encouragement and support. When I arrived at NASA/JPL, all I felt was gratitude for the opportunity to work at this storied center of robotic space exploration. One of my earliest experiences at JPL was seeing the effort made to share our work through the yearly "open house," the public lecture series, and tours about our space missions. I was struck by the responsibility we have as engineers and scientists to share our results and explorations with the public that funds NASA.

To do my share, I began lecturing for museums and schools in my spare time after receiving some excellent mentoring in effective outreach. What I was unprepared for was the response of the young women (and the young men) in the audiences. I found that kids of all ages were intensely captivated by the stories of the missions. I heard from so many young women after these lectures that wanted advice and encouragement on how to get involved in Aerospace. There have been stages in my career where I have considered taking a break from outreach due to the time commitments involved. Yet I have continued. I am here because of my mother and her encouragement. Staying the course in mentoring through outreach over the decades is how I add my voice to the support that other young girls and young women can feel about being a future engineer.

Avoiding a Stall

Have you ever taken a role you were not excited about but had to show you were a "team player"? What was the outcome?

Robotic space exploration, especially deep space missions, can have a hard deadline in terms of interplanetary launch windows. If one of our rovers to Mars is not ready for launch, the next opportunity can be more than two years in the future. Thus, preparing a spacecraft for launch is an "all-hands-on-deck" endeavour. There have been times when I have been working on other missions and the call has come to join a different mission team because of my prior positions. Experience in Mars surface operations is a very specific skillset with a limited pool of engineers to pull from.

In one case, I was working on an asteroid mission and the next rover mission needed leadership in how we were going to conduct training for accelerated operations on Mars. From my combined time in military, commercial, and NASA space operations, I had the right background for that position even though I was happily engaged elsewhere. It was clear this was a special case and a special need and one with the launch date fast approaching. I accepted the position and plunged right into the challenge.

We all get asked to do roles for the good of the organization. Being flexible about saying "yes" opened different doors that I could not have anticipated, and I was glad to re-join my colleagues on the Mars missions again. It sounds like a cliché to say that in every opportunity there is another opportunity, but it is often true.

Powering On

Was there a significant event that changed your career trajectory and what was it?

A pivotal event in my career trajectory happened when I was stationed at US Space Command at Cheyenne Mountain Air Force Station in Colorado Springs. I was busily working at my console and eavesdropping on colleagues behind me. US Space Command was a "purple assignment"—meaning we served in our duty positions with members from different branches of the military. Thus, my colleagues were all from varying backgrounds, and they were discussing military space operations with great passion.

I completely resonate with the "duty, honor, country" side of the Air Force, and I am so glad to have served. However, at that moment, hearing the excitement in their voices, I was reminded clearly that I had not changed. As much as I enjoyed the USAF, my passion for space exploration was still what it was when I was a teenager. My goal was NASA/JPL and the robotic exploration of the solar system and beyond. It was time for me to take the risk of leaving a secure job, and what would likely have been a rewarding and wonderful career in the Air Force, for the possibility of getting my dream job at NASA/JPL. I did make it to JPL less than one year later, and it is still my dream job.

Kirsten Dreggors

Vice President, Engineering
Northrop Grumman Aeronautics Systems

About the Author

Kirsten Dreggors is Vice President of Engineering and leader for the Manned Aircraft Design (MAD) Center of Excellence (CoE) at Northrop Grumman Aeronautics Systems, a premier provider of military aircraft, autonomous systems, aerospace structures and next-generation solutions.

In this role, she leads more than 2,600 engineers on E-2D Advanced Hawkeye, Joint STARS, Airborne Laser Mine Detection System, and restricted programs. She also leads the MAD CoE of nearly 5,000 employees at the Melbourne, Florida site, overseeing the continuing growth of staff and facilities and serving as Northrop Grumman's Lead Executive in the community.

Most recently, Kirsten was the Director of Vehicle Engineering responsible for leadership of the MAD CoE vehicle engineering organization and the vehicle engineering Community of Practice for the sector. Since joining Northrop Grumman in 1997 as a Systems Engineer, Kirsten has held positions of increasing responsibility and leadership across a variety of programs including Joint STARS, E-10A, Firebird, Global Hawk, F-35, and Fire Scout. She has led integrated product teams, overseeing the engineering development, design, integration and test of mission systems and payloads, and contributed to the success of manned and unmanned platforms at all phases of the program life cycle.

Kirsten is active in Northrop Grumman's leadership development programs. A graduate of the Northrop Grumman Corporate Engineering Council mentoring program, she serves as a mentor and resource for the Women in Leadership and LEADING programs. Kirsten earned a Bachelor's degree in Aerospace Engineering and a Master's degree in Mechanical Engineering from the University of Central Florida.

Questions and Answers

Personal Career Insight

What inspired you to choose a career in Aerospace and what has given you the most satisfaction during your career?

I became fascinated with airplanes in junior high and set my sights on building the world's fastest airplane. I was fortunate to have a wonderfully supportive mother who took me to the library to research careers in aviation. From that time on, I knew I wanted to be an aerospace engineer, and I steered myself towards technical and design coursework. My first-ever flight in an airplane wasn't until the end of undergrad; my aerodynamics instructor took several of us in his four-seater Cessna just before graduation. It was awesome to finally experience flight after years of studying it in the classroom. I couldn't wait to get started.

I haven't worked on the world's fastest airplane—yet!—but a highlight of my career was a project to design and build the first Firebird demonstrator aircraft in a single year. A group of Northrop Grumman colleagues partnered with a team from Scaled Composites in the Mojave Desert to tackle the challenge. The Firebird is an optionally piloted, medium-altitude surveillance aircraft. This was in 2009-2010 and we were really delving into new territory with the manned and unmanned design of the aircraft. We literally sketched the airplane on a cocktail napkin, and exactly one year later it took to the sky for the first time. It was an amazing experience on so many levels. We were doing Agile before it had a name; it was just an iterative design process and we made adjustments as needed to meet the mission requirements. What we were able to accomplish in a single year was incredible, and I'm very proud to have been part of such an amazing team.

Work/Life Balance

Did you ever have to make a move to advance your career (within your company or changing companies) that impacted your family life, and how did you balance the two?

I launched my career with Northrop Grumman in Melbourne, Florida. I was hired as an engineer responsible for technical drawings on the Joint STARS program. A few years and a few promotions later, I was offered an assignment on a different program across the country in San Diego. It was a lateral move and felt like a bit of a risk, but I saw the potential to gain new experience and develop contacts within the organization that would position me for a future career move. I'm so thankful I took a chance on myself because it really paid off. Not only did I broaden my skillset and experience, I also immersed myself in a different workplace culture, which gave me insight into different thought processes, priorities, and how decisions get made. That organizational intelligence is hard to quantify when you're considering a lateral career move, but it has served me well ever since. In addition, I was able to parlay that experience into a promotion while I was working on the West Coast.

From a family perspective, it was a tough decision. My husband—who is also an engineer and works for Northrop Grumman—and I had many long conversations and did plenty of soul-searching to determine whether this was the right thing for us. I moved to California first, and he found a job and transferred with me about six months later. Moving your spouse for a job is a big deal, and for us, the key has been open and honest communication, mutual respect for one another's career ambitions, and always making decisions together.

Mentorship/Sponsorship

How important was mentorship/sponsorship for your career? Have they been men or women? How was the relationship established?

I've been fortunate to have some outstanding mentors and sponsors—both men and women—over the course of my career. I've approached colleagues and leaders for guidance and mentoring, both through formal programs and informally. Sponsors are different from mentors: they're earned. A sponsor is a leader who recognizes your performance, achievements, and aptitude and is willing to put their reputation on the line to help advance your career. A leader in your organization may take notice of a project you work on early in your career; it might seem insignificant at the time, but those first impressions and relationships can open doors for you years down the road.

For me, returning to Florida stands out as an example of why having a sponsor is so important. Transferring to California was a great career move for me, but Florida was home and my husband and I knew we wanted to get back at some point. I credit one sponsor in particular for helping me make that happen. When I heard that the Director of Vehicle Engineering at the Florida site was planning to retire, I reached out to my sponsor to discuss my fitness for the role. Early in my career, my sponsor took notice of my willingness to raise my hand and take on special assignments—working in Mojave on the Firebird program is one example—and recognized the value I brought to each program I supported. I had earned the bona fides to move into a director-level role, and I had demonstrated my leadership skills to other leaders. Once again, I transferred across the country, but this time it was for a promotion.

A career doesn't happen in a vacuum. It takes hard work and commitment on your part, but oftentimes we need the help of influential leaders in our organization to take advantage of big opportunities. A sponsor has a seat at the table where you don't, or has the ear of other leaders you don't, and they can help your advance your career beyond your achievements and immediate network.

Avoiding a Stall

How did you develop organizational savvy? And how did it help your career?

Knowing your organization is foundational for success. All the technical skills in the world won't necessarily take you the entire way. You need to understand how your organization makes decisions based on what's important, and who makes them based on leadership dynamics. That's where your professional network comes in. For me, being able to move around the organization was a tremendous benefit. I've made a conscious effort over the years to keep in touch with leaders and peers I've connected with. We may live in different states, but whenever our paths cross on a business trip, we make a point to meet for dinner or coffee. With many of these relationships, we may not talk for months or even a year, but like old friends who know each other well, we pick up right where we left off.

Northrop Grumman has great programs to facilitate the development of strong networks for both early-career women and women in leadership. I was in the first cohort of the Women in Science and Engineering (WISE) Program. The company developed the WISE Program to support young women engineers by facilitating mentorship and providing tools to help us craft a career path. Many of the women I met through that program are also still with the company, and to this day we help one another. Sometimes, you just need a trusted confidant to bounce a new idea off before presenting it to your team or boss. Some have technical or functional leadership roles in the organization I now lead, and I am so grateful for the expertise they bring. I'm involved in the WISE Program today, but on the other side as a coach and mentor. I was a beneficiary of that program and I see the value it delivers. I want my success to inspire and enable the next generation of young women engineers to pursue their career goals here too.

There's also a Women in Leadership program designed to give continuing support to women as they step into leadership. Shifting from an individual contributor role to a management role brings a new set of responsibilities and expectations. There's more at stake, and having seasoned women leaders to guide me as I came up the ranks not only kept me going but gave me valuable advice about navigating the organization as a leader.

Powering On

As you hit career obstacles, what motivated you to keep going?

It's almost inevitable that at some point in the course of a career, everyone will feel stuck. The key is to use that as an opportunity for growth rather than letting it frustrate or demotivate you. When you're feeling stuck is precisely the time to look for other ways you can contribute - maybe it means finding a mentor to help broaden your perspective or setting a stretch objective that keeps you learning. Try to find new ways to contribute or otherwise stay motivated.

It's also important to establish something for yourself outside of work that keeps you motivated. It could be a hobby, your family, or a civic- or faith-based organization. In between the good days, and climbing the corporate ladder, there will be rough days or sometimes a rough month, and you need other sources of satisfaction and accomplishment outside the office.

My husband and I enjoy spending time off work outside in plenty of fresh air. We own a Jeep and two motorcycles. As I've climbed to higher levels of leadership, I am more removed from the technical engineering work on a daily basis. Tinkering on the Jeep and the bikes gives me an outlet for my technical engineering brain. It's satisfying to be able to work with my hands and then take a fun cruise up the beach enjoying the beautiful Florida weather.

Kathryn Elliott

Defense Sector Capability Manager
Performance and Aerothermal Systems
Rolls-Royce

About the Author

Kathryn Elliott, the Defense Sector Capability Manager for Performance and Aerothermal Systems at Rolls-Royce Corporation, holds a Bachelor of Science degree in Mechanical Engineering from the University of Michigan and has over 30 years of experience in Aerospace focusing on gas turbine engine thermodynamic cycle design and analysis, and component aero design, test, and aftermarket data analysis. She is currently the Rolls-Royce Global Lead Subject Matter Expert for In-Service Performance. She is Chair of the SAE International E-32 Committee, which standardizes the industry best practice for aerospace propulsion systems health management, and she serves on the Management Team for the Prognostics and Health Management Society.

She sits on the University of Michigan's Industry Advisory Board for the Aerospace Engineering Department and has been an invited speaker at the Society of Women Engineers (SWE) National Conference in 2013 and 2018. In 2014, Kathy led the launch of the first Rolls-Royce North America employee resource group, Rolls-Royce Women North America, and has since fostered seven additional Employee Resource Groups (ERG). She was recognized with Rolls-Royce Engineering's "Inspirational Leadership" Award for 2016. She has served as Mentor and Judge for FIRST Indiana Robotics events for the past five years, particularly for the all-girls off-season event Raising Awareness of Girls in Engineering (Indy RAGE).

Kathy is passionate about science, technology, engineering, and mathematics (STEM) outreach for girls and redefining the journey to success for women. Along her career path, she has experienced many different work styles/lifestyles: full time, no time, stay-at-home mom, part-time and full-time working mom, contractor, working from home (decades before Covid-19!), and as a sole proprietor—all while functioning as an engineer in the aerospace industry.

Questions and Answers

Personal Career Insight

How did you decide between a leadership vs. technical career track?

I had few STEM role models growing up—my father was a minister, my only uncle was a minister, and my mom and only aunt were minister's wives. The first "role models" I remember were Diane Fossey and Jane Goodall, who in my little girl's mind had the perfect jobs: sitting alone in the jungles of Africa watching apes and chimpanzees. Three things appealed to me about that as a career: (1) the science of animal behaviorism, (2) the opportunity to learn things that no one had ever learned before, and (3) the absence of all human interaction. Those three elements, science, discovery, and seclusion, eventually set the stage for my expectations in an engineering career as well.

As an introvert, I much preferred to spend my time coding and analyzing data in the chase for turbine engine efficiency to dealing with the drama that working with people inevitably brings. Even so, one of my early managers noted that I had the potential for management and tagged me as one to watch. My progression to management, however, was deferred due to circumstances: working in Defense following the end of the Cold War meant a layoff. Then I started having children and began to work part time. I spent 10 years on the classic "mommy-track" with no promotions. So it happened that I enjoyed a rather long time (25 years) focused on a technical career track, and altogether ignoring office politics.

Over the course of that time, however, the level of design integration increased significantly as advanced tools and software allowed systems to be optimized across a much broader landscape. Ironically, this increasing technical complexity drove a larger need to focus on the people and their human connections to achieve the best solutions. It became clear that I could make a larger contribution as a leader than I could as an individual specialist. I had also changed. I had studied motivation and enjoyed connecting with others and building collaborative teams.

Eventually, the move to management and on to leadership felt very natural and brought a fresh set of challenges. I get my biggest satisfaction in helping engineers find their groove and achieve things they didn't dream possible. It's exhilarating to influence the technical direction of new technologies emerging in aerospace and to be a part of building new capability.

Work/Life Balance

Did you ever have to make a move to advance your career (within your company or changing companies) that impacted your family life, and how did you balance the two?

My husband and I had three children in five years, and I worked part time for the first ten years of being a parent. This was unheard of in the 90s, and in fact, my company didn't know how to do "part-time" and so I had to work as a contractor. Those years were fabulous—there were no promotions and few raises, but I enjoyed the time spent with my children and being involved in their school and activities.

Meanwhile, my husband worked full time, and we continued to live in his hometown with his extended family around us. When our youngest was old enough to go to school full time, we took stock and decided that since I had the higher-earning power, it was time to leverage that potential as we had three college funds to work on! My search for full-time work resulted in our relocating 250 miles away to a new city, with new schools, new church, new friends, and no family. It also meant my husband had to leave his job. So we did a swap—I worked full time and he stayed home with the kids! He was the only stay-at-home dad in the neighborhood. It took about a year to adjust, but we settled in, and all made new friends in our new town.

Kathryn Elliott

The children were young enough to adjust fairly easily, and after six months, my husband got a part-time job working at home. The parental role-swap had a considerable effect on all our relationships—the children grew closer to their dad and we each developed a greater appreciation for the working spouse versus the stay-at-home spouse. Everyone now agrees that it was a good move at the right time for us.

I should note that the downside of working part time while the children were young was that I was on a career plateau and went 10 years with no progression. This was a trade I was willing to make for the privilege of raising my children. When I reentered the workforce full time, I was clearly perceived as a risk—Is she going to stay? Can she hack full time? Is she serious about her career? It took nearly another five years to reestablish my credibility and put those doubts to rest. I still have no regrets about the choice I made and consider myself fortunate to have had the options I did. I'd like all women to have the freedom to choose the right path for them and their families.

Mentorship/Sponsorship

What activities have you engaged in that have helped other women achieve success in their aerospace careers?

In college I enjoyed the benefit of an excellent study group. This was an informal group of 10 men that formed up to work on homework and study for tests together. The group was well balanced and the guys were smart and fun and treated me with respect as an equal. I was the only girl in my class, so this group was a godsend. We studied together for all four years and are still good friends today. The success of this group led me to assume that my peers in the working world would behave in a similar fashion and gender equality would be achieved as the old guard retired and left the workforce. How naïve was I!?

Early in my first job, I presented a technically detailed review to several top military brass that went very well. I was so excited that I called my mother that night to tell her about it. She asked me, "Were you the only woman in the room?" I proudly told her that "Yes, I was the only woman in the room." I was pleased that there WAS a woman in the room! Fast-forward 30 years and I hear my mother's voice in my head ask the same question, except now she says, "Are you STILL the only woman in the room?" My answer is again, yes, but now I'm not so proud of it. This lit an absolute fire in me and I set out to change that answer.

I engaged with the Society of Women Engineers (SWE) on several levels. I recruited for Rolls-Royce at the national conference and was selected as a conference speaker for several years. I joined the local professional section in Central Indiana and participated in events mentoring local women. I joined in women's initiatives at my alma mater, mentoring and speaking with the University of Michigan SWE section and the Women in Aerospace and Aeronautics. As an Industrial Advisory Board Member for the Aerospace Engineering Department, I've been part of a rumbling change to bring Diversity, Equity, and Inclusion to the forefront of priorities for the department.

As Rolls-Royce Corporation began to recognize the power of diversity and inclusion, I launched the first Employee Resource Group (ERG), Rolls-Royce Women in North America, and served as its president for three years. This ERG is engaged all across Indiana, partnering with Women & Hi Tech, Indy Women in Tech, and Indy Women's Forum, for the professional development of women and STEM outreach to underserved girls. One of my favorite events is the FIRST Robotics RAGE event (Raising Awareness for Girls in Engineering). This annual robotics competition encourages girls on their high school robotics teams to take on all the roles, from driver to mechanic, in a full day of competition with the boys in the stands cheering them on!

The myth of gender-equality-by-evolution is still very prevalent and is often used as a cop-out for those who refuse to acknowledge the problem and don't want to do anything to change the status quo. I aim to be intentional and deliberate about accelerating the pace of change for women, I only wish I'd started sooner.

Avoiding a Stall

Were you ever presented an "opportunity" you declined and did it hurt your career? If not, how did you overcome any negative impact?

At one point in my career, I led a team focused on the engine health management for a large commercial fleet with many different airline customers. It was an exciting time of rapid capability growth, with opportunities for technical innovation and close collaboration with the airline customers to deliver improvements. I enjoyed the challenges of this role very much and achieved several notable successes.

Those successes generated visibility for me, which resulted in several opportunities to move into larger roles in other areas of the company. I was advised to ride the wave of success and "strike while the iron is hot". I resisted though because my current job was absolutely the best one I'd ever had and I wasn't quite done enjoying it yet. I was at a peak and wanted to dwell there just a bit longer. I continued in that role for another year, and in that time, I felt that the peak had definitely passed and I was finally ready for a new challenge. Fortunately, a suitably challenging role did materialize that matched my skills and aspirations and I was able to embark in a new direction and take on more leadership accountability in the organization.

In retrospect though, I took a big risk by delaying a move and turning down opportunities. I was very lucky to have a second chance, and it might not have worked out so well. In letting opportunities go by while I continued to "enjoy my job," there was a significant chance that I could have been stuck in a job that became increasingly less rewarding. Now I advise others to leave a job at the peak or even just before the peak. Ride that wave of success to launch into your next role. Think of it as not overstaying your welcome and leaving with them wanting just a little bit more rather than getting too comfortable or stale.

Powering On

As you hit career obstacles, what motivated you to keep going?

I've had my share of various obstacles, but as I've navigated over, around, and through them, I realize that they really do make you stronger. My first major obstacle was getting caught up in a company reduction about eight years into my first job. Although I'd seen it coming at least a year earlier and my colleagues expected it, when it happened it was still quite a blow to my confidence and led me to question my career path. With my husband's support, I began to explore other opportunities for my future.

Based on my hobbies and interests, I worked for a time as a music teacher, a preschool teacher, and then for a tech start-up (not common in 1992!). While I enjoyed these activities for a time, I came to realize that they were more hobbies and not work I wanted to pursue long term. When I returned to a role in aerospace engineering, it strongly reaffirmed that I was meant to be in engineering. This solid knowledge has motivated me and kept me afloat throughout my career when the road got rocky.

One of those low times was following a massive company reorganization when I found myself grieving and depressed. It feels a bit embarrassing to say that I had so little resilience, but it was true—everything I had built and worked to achieve was shattered and I was depressed for nearly a year. I found it very difficult to remain motivated and focused. The key for me was focusing on others who were depending on me and looking for things to celebrate. That took the form of writing nominations for the local Leading Light Awards. These awards celebrate women in STEM in Indiana and are celebrated every two years. I spent many weeks interviewing amazing women and listening to their stories so I could write the best possible nominations. Focusing on their success was extremely therapeutic for me and helped pull me out of that funk and be the leader I needed to be. It's true that in the face of adversity, the best advice is to pull yourself up, look forward, and march on.

Kathryn Elliott

Carol Erikson

Vice President
Systems Engineering and Digital
Transformation
Northrop Grumman Space Systems

About the Author

Carol Erikson is Vice President of Systems Engineering and Digital Transformation for Northrop Grumman Space Systems sector, a space and launch systems provider serving national security, civil, and commercial customers.

In this role, Erikson is responsible for leading systems engineering and digital transformation initiatives across the sector that will enable Northrop Grumman Space Systems team members to leverage Model-Based Systems Engineering and Integrated Digital Environment and other transformational capabilities to ensure effective execution across the full program life cycle from architecture and concept studies through program execution, product delivery, and sustainment.

Most recently, Erikson served as Vice President and Enterprise Program Manager for the Ground-Based Strategic Deterrent program, where she established and led the nationwide team responsible for capturing and executing the Technical Maturation and Risk Reduction program and for developing innovative engineering and digital environment solutions critical to capturing and preparing for the Engineering, Manufacturing, and Development program.

Erikson joined Northrop Grumman in 1987 as a systems engineer and has technical, supply chain, functional management, and program management experience in space systems. She served as the Director for Mission 1 System Enhancement and Captures, Deputy Program Manager for Program Integration on the Advanced Missions Programs portfolio, and Project Manager for payloads for the National Polar-orbiting Operational Environmental Satellite System. In 2013, Erikson was appointed Vice President of Engineering for the heritage Northrop Grumman Aerospace Systems sector's Space Systems Division, where she led engineering efforts across the Space Systems portfolio, including restricted space, the James Webb Space Telescope, and the Advanced Extremely High-Frequency programs.

Ms. Erikson earned a Bachelor's degree in Engineering and Computer Science from Stonehill College and a Bachelor's degree in Electrical Engineering from the University of Notre Dame. She also earned a Master of Science degree in Electrical Engineering from the University of Southern California and completed the Executive Management Program at the University of California Los Angeles.

Questions and Answers

Personal Career Insight

What inspired you to choose a career in Aerospace and what has given you the most satisfaction during your career?

I was inspired to choose a career in Aerospace by the passion and excitement that a female systems engineer demonstrated when I interviewed for an entry-level position at TRW. I was a senior studying Electrical Engineering at the University of Notre Dame in the winter of 1987. The company was very smart about their recruiting plans, sending me from South Bend, IN to Redondo Beach, CA in February. I certainly fell in love very quickly with the weather in California! But the real turning point for me was the interview that I had with Debbie Fitzgerald Simmons.

At the time, she was the lead systems engineer for NASA's Gamma Ray Observatory. She was so passionate for the work, the importance of the mission, and the quality of the people on her team. She left a great impression on me and really helped TRW stand above my other options. As graduation day grew closer, I found myself with four compelling job offers—working on satellite flight software at TRW in California; working for GE Defense Systems in either Pittsfield, MA or Rochester, NY (talk about a contrast in weather from California!); or working in factory automation at a GE factory in Connecticut.

While the pull to move back to the East Coast to be close to my family was tremendous, ultimately the opportunity to move to the West Coast, learn about satellites, and experience the passion and commitment that Debbie had demonstrated was a bigger pull. I thought I would move to California for a year or two to experience the aerospace industry and ultimately move back to the East Coast in another technical field. That was 33 years ago and I am still here!

Ironically, it is that same commitment to the importance of our missions and recognizing the quality of the people that has brought me the most satisfaction during my career. Over the course of my career, I have had the opportunity to work on projects that have been of tremendous importance to our nation. I know that what I am working on every day is critically important to our nation whether the mission is helping to understand the origins of the universe, on programs like the Chandra X-Ray Observatory, or to preserve freedom, on missile defense satellites, national weather satellites, intercontinental ballistic missile programs, and many restricted programs of critical importance to our nation's (and the world's) security.

These systems are complex and challenging to design, produce, and operate, which leads to one of the other most satisfying aspects, the ability to learn and grow. I have literally learned something new every day over the course of my 33-year career. And, of course, none of this would be satisfying without the tremendous people who have formed my "work family" over the past three decades. TRW (now Northrop Grumman after an acquisition in 2002) attracts the highest quality of people—a diverse team of the most intelligent people you will ever meet committed to the importance of our missions and willing to work together like a family. It is this combination—the importance of our missions, the quality of our people, and the ability to learn and grow every day—that has brought me the most satisfaction from my career.

Work/Life Balance

Did the pursuit of a career impact your decision whether or when to have a family? What was the impact of that decision?

The pursuit of my career impacted the timing of my marriage and the ultimate blessing of raising a family. But it was less of a decision and more a matter of fate and the impact of choices throughout my life. I have always been someone who has valued diverse experiences and living life to the fullest. While I was in my 20s and 30s, I was very committed to my career. The nature of our work often requires long hours, weekend work, and extensive travel. With my passion and commitment to our missions, I rarely said no when work demands called. But I was never a person who believed in all work and no play. On any given weekend, I could be found skiing, mountain bike riding, running, surfing or playing soccer or softball. I also enjoyed traveling around the world both for work and vacations.

As you could imagine, finding time for a committed relationship was difficult given such a busy life. But, even more fundamental, was the ability to find the right partner in life. During this time, I had several long-term relationships, but I found it very difficult to find someone willing to put up with my crazy ways. In my late 30s, just a few weeks after my mother passed away, I had the opportunity to play in a volleyball tournament with Joaquin Hernandez. From the start of that tournament, the conversations with Joaquin were different than anyone I had met before. He was supportive of me both on and off the court. He cheered me on over the next few weeks as I trained to climb Mt. Whitney, the highest mountain in the continental United States. And, since he was working for a non-profit that sold training to the aerospace industry, he recognized the importance of my work and what it meant to me and to the nation. I often think my mother had to go to heaven and then send the right partner for life my way! We got married when I was 40, and I celebrated my 41st birthday during a safari on our honeymoon in South Africa.

We knew from the start of our marriage that we wanted to have children but we never knew how hard that would be given our ages! When I was finally pregnant with our first child, my doctor would remind me that I was 44 going on 45 (as if

I needed a reminder!). We were blessed with our first son, Ricky, in November 2009. At the time I was a Program Director and I remember being very nervous about the impact having a child would have on my career. Although the initial transition back to work was hard, ultimately I was able to balance work and family with a lot of help from Joaquin, the benefits of our on-site daycare center, and the support of my manager and fellow team members. Our next blessing came in July 2013 when my daughter Elizabeth was born. I will always remember that month, not only because it was when Elizabeth was born but also because I became a Vice President!

Carol Erikson

So, while fate and personal choices, both in my career and outside of work, may have led me to getting married and having children in my 40s, I would not change a thing. Now, in my mid-50s, I have the joy of raising two beautiful children, the happiness of a marriage with an amazing husband, and a rewarding career as a Vice President at Northrop Grumman. This journey has worked well for me and I would not change a thing!

Carol Erikson

Mentorship/Sponsorship

How important was mentorship/sponsorship for your career? Have they been men or women? How was the relationship established?

Mentorship and sponsorship have been extremely important to my career. My first mentor was Sue Stavlo who was assigned as my mentor. I had experience working directly with Sue before, so I knew that I could benefit from her years of experience working on flight software. But, it was her willingness to provide honest feedback that would lead to the most beneficial mentoring that I could receive. Eventually, as my career continued to progress, Sue advised me to reach out to her husband John for additional mentoring. John had started at the company as an engineer and worked his way up to Vice President. He was able to share advice on how to successfully advance through the corporate structure as well as how to address some of the unique challenges associated with leadership roles. Through the years, I became great friends with both Sue and John and received some of my best mentoring and advice as we skied at Mammoth Mountain.

As my career progressed, I have relied on both men and women as mentors. The most impactful have been Bob Burke and Peggy Nelson. In both cases, the relationship started as I worked directly for them, but they each had an impact that would expand far beyond the years that I worked for them. Bob was one of best leaders that I have ever met. He invested time in me and pushed me outside of my comfort zones. I was lucky enough to begin working for Bob and learning from him just as his career path was also accelerating. So, as he was pulled into roles of higher responsibility, I had the opportunity to follow and take on the roles that he was leaving behind. In each case, he ensured that I had the support that I needed to succeed. Over the years our career paths would diverge, but I still leveraged him as a valuable mentor when I faced new challenges.

The first experience that I had with Peggy as a mentor may not have ended as she had hoped. She had asked me to move into a management role reporting directly to her. At the time, I was in a management role on a different program, and despite the significant urging from Peggy, I was unwilling to make the move. Luckily, Peggy did not hold that against me! I asked her to keep me in mind for future roles, and about a year later that opportunity came my way. That move started an eight-year journey where I worked directly for one of the best program managers in the industry. Peggy shared a similar quality with Sue, an amazing technical depth combined with a willingness to tell me exactly what was on her mind. I learned so much from her direct feedback and the way she worked. Peggy also demonstrated to me firsthand how to successfully balance family and work (or as she put it enjoy the "work-life see-saw"). And, ultimately, Peggy became my biggest advocate as she opened the door for me to become a Vice President.

I can honestly say that mentorship and sponsorship throughout my career have helped shape who I am today. I have valued each piece of advice I have received, even if not always following the advice, and plan on continuing to leverage mentors as I move forward with my career.

Avoiding a Stall

Were you ever presented an "opportunity" you declined and did it hurt your career? If not, how did you overcome any negative impact?

I set my goal of becoming a Vice President at TRW/Northrop Grumman early in my career after seeing the impact that leaders like John Stavlo, Dick Croxall, and Dale Hoffman could have in the role. In 2008, after 21 years at the company, I finally had that opportunity. I was asked if I would be interested in a Vice President and Program Manager role—my dream job!

The only problem was the position was located in McLean, VA. Joaquin and I had married in 2005. In 2007, he had started his own company and had spent a year building his clientele. So, needless to say, it was not an ideal time for him to consider a move to the East Coast. I did what any good engineer would do and built an Excel spreadsheet to help us through the analysis of the pros and cons. We spent hours talking about the potential of moving to Virginia to support my dream job and even considered the potential of becoming a bi-coastal couple. However, there was one big factor that weighed in the decision and neither of us had complete ability to control that factor or the outcome—our desire to start a family. We had started on that journey soon after getting married and found the challenges that many face in a similar situation if they try to have children in their 40s.

It seemed like I was faced with a clear decision point: pursue my dream job in Virginia and give up on my dream to have children or stay in California with the hope that we would be blessed with children and risk not getting another chance for a dream role. As my luck would have it, I had a meeting with our CEO early Monday morning and would have to tell him that I had decided not to take the job! I was so nervous to deliver the message but then so relieved when I heard his response: "Carol, I understand, everyone needs balance at some time."

The years went on and I continued to be challenged in director-level roles in California and continued to learn and grow every day on the job. And, of course, the ultimate blessing came in 2009 with the birth of Ricky. So I never regretted the decision. However, after a few more years had passed I began to wonder, "Had I missed my chance for a Vice President role? Was that the only opportunity that would come my way?" Those concerns were put aside as I became pregnant with Elizabeth in 2013, and that's when I got the call from Peggy about the potential for a Vice President role.

Ultimately, passing on the first Vice President position may have changed both my career path and my family life, but it is a decision I will never regret. Following that decision, I have had the opportunity to experience the blessings of motherhood while also reaching my dream roles at work and experiencing both the challenges and rewards associated with executive leadership in the aerospace industry.

Powering On

Was there a significant event that changed your career trajectory and what was it?

The significant event that changed my career trajectory was the decision to move from a flight software department to systems engineering. When I made this change my passions and capabilities aligned with the needs at work, and my career progression really began to accelerate. (This was also the time that I started working for Bob Burke who I mentioned above, and I benefited directly from his mentorship as my career progressed directly behind his).

Although the early years of my career were in-flight software, I had never developed a line of code at work. I spent my time developing requirements and designing and testing hardware to software interfaces—all key aspects of systems engineering. I also found that what I enjoyed about flight software was learning about the rest of the system. The flight software had to control the attitude of the spacecraft, the temperatures of the equipment, and ensure that commands and data could be transmitted to and from the control stations on the ground. I enjoyed my interactions with the engineers in all of these areas and learned about their unique parts of the system. It was this attraction to understanding the bigger picture that drove me towards systems engineering.

It is also interesting to note that in my early career days, I spent most of my time outside of work planning and organizing large "teams." Whether it was a softball or soccer team or a large weekend ski trip for all my friends or a backpacking trip around Greece, I gladly took on the role of the "planner" and de facto leader. As I moved into systems engineering, and my technical depth and capabilities were aligned with the team's needs, I was also asked to take on leadership roles. I had found my sweet spot both technically and organizationally.

The roles that I had at this time, supporting the launch and operations of our satellites, also provided foundational capabilities that would be key to my success in future roles. These roles provided a unique opportunity to work directly with our customers and understand the true mission for our system—how the systems were used in day-to-day operations to support our nation. I began to understand the importance of mission systems engineering and ensuring our systems are designed to meet the needs of the end users.

Also, when a problem happened on these systems, whether it be on the launch pad or while in orbit, you had to very quickly work with the subject matter experts who had designed the system, determine the root cause, or what may have caused the issue, develop a workaround or contingency plan, and either launch the satellite or bring it back into operations as soon as it was safe to do so. These capabilities to form multi-disciplinary teams, quickly solve issues, evaluate alternatives, and implement quick fixes would provide me a toolset of technical and leadership skills that were foundational for my success in all subsequent roles as a Program Manager and Vice President.

Moving from the flight software department to systems engineering changed my career by aligning my skills and passions with the program needs, expanding my technical expertise and knowledge of the importance of mission systems engineering, and building my leadership toolset with fundamental capabilities needed to lead teams and solve hard problems.

Kim Ernzen

Vice President, Naval Power
Raytheon Missiles and Defense

About the Author

Kim Ernzen is Vice President of Naval Power at Raytheon Missiles and Defense, a business of Raytheon Technologies. Ernzen's team provides the United States and its global allies in the maritime domain with sensors, command and control, and precision weapon solutions to protect allied ships and sailors around the world.

Ernzen served as Vice President of Air Warfare Systems at Raytheon Missiles Systems, prior to Raytheon Company's merger with United Technologies Corporation in 2020. Before that, she served as Vice President of the Land Warfare Systems product line and as Vice President of Operations for the business.

Ernzen currently serves as Board Member for Tucson Values Teachers and previously served on both the Wichita Children's Home Board of Directors and the Junior League of Wichita Board.

She holds a Master's degree in Aeronautical Engineering from Wichita State University, as well as an executive MBA and Bachelor's degree in Aeronautical Engineering, also from Wichita State University.

Questions and Answers

Personal Career Insight

What inspired you to choose a career in Aerospace and what has given you the most satisfaction during your career?

How did you decide between a leadership vs. technical career track?

For me this decision came early in my career. I initially started my career in engineering after graduating with my BS in Aeronautical Engineering. Growing up I always had a love for math and science. In junior high, I became interested in airplanes, trying to understand how they worked, what made them fly, and so during high school, my counselor suggested I study engineering in college. I honestly had no idea when I decided to choose that major that I would be in the minority for years to come.

After working in the General Aviation industry in Wichita for a couple of years, I decided to go back to school, while still working, to obtain my Master's degree in Aeronautical Engineering. During this same time period, I was working as an analyst in the aero group, creating performance charts for a re-engine effort, when I realized I wanted to consider moving into more leadership-type roles. While I enjoyed the technical aspects and solving the complex problems, I really started to see my passion was bringing teams together to solve the complex problems of both technical issues as well as satisfying the customer and growing your business.

Initially, I found my first leadership role still within the Engineering organization as a Project Engineer/Chief Engineer, but had my first opportunity to oversee a larger team as well as start to gain exposure to the financial side of the business. I also decided at this time to go back to school one more time to get my MBA, as I felt I needed to develop my business acumen. For me personally, it was the best decision I could have made as it allowed me to not only learn the business side, but the experience back at school with working professionals helped me learn the business challenges other industries face and the strategies they implement to overcome their hurdles.

Looking back over my career, I am so glad I had the technical foundation from which to pivot from. Given that we work in an industry with cutting-edge technology, I believe it has allowed me to be a more effective leader while overseeing the complex programs and products we deliver to our customers.

Work/Life Balance

Did you ever have to make a move to advance your career (within your company or changing companies) that impacted your family life, and how did you balance the two?

Yes, I have made a couple of moves to advance my career, and they have been a mix of within my company and changing companies. The first move came when my two daughters were still relatively young (fourth grade and Pre-K), and I had the opportunity to relocate with a lateral position. The company I was with had been a part of the company I was returning to; however, the division had been divested about two years prior. At the time, both my husband and I were working at the divested division and decided we wanted to get back to the parent company so that both of us would have more potential career opportunities. I was the one that received the initial offer, and so we decided as a family to make the move to in essence restart our careers.

Kim Ernzen

The second move would come nearly five years later, and this time it would be within the company. I was given my first role as Vice President but found us once again uprooting our family only this time our daughters were in eighth and fourth grade. It was also at the point when we decided as a family that I was the one who really wanted the career, and so my husband decided to stay at home and be the parent really managing our family and home life. I feel incredibly blessed to have a spouse who completely supports me in my decision to want to fulfill my career dreams, while also helping me to be the kind of mother I want to be. It certainly is not always easy, I rarely know who their teachers are but I work my schedule to make sure I do not miss school or extracurricular events. I also try to be the inspiration for my daughters, demonstrating that they can be whatever they choose to be and that, yes, there will be sacrifices, but with hard work, shooting for your dreams you really can have it all.

Mentorship/Sponsorship

How important was mentorship/sponsorship for your career? Have they been men or women? How was the relationship established?

Both mentors and sponsors are critical to your career and professional development. I have been fortunate over my career to have both, some of whom were instrumental in getting me on the trajectory I am on today. Fortunately, or unfortunately, depending on how you look at it, all of my mentors/sponsors have been men. To date, I have never had a female supervisor, which may be similar for some of you.

Early on in my career, I tended to have only mentors, and typically the relationship was started by me reaching out to them. Usually, it was someone either recommended to me by my supervisor or someone I felt I could learn from. What I think is important is to have an array of mentors, as each of them will provide a different view and input to you. Depending on what specifically you are looking for; in some cases it may be to have someone serve as a safe sounding board for you. Other times it might be someone you want to learn specific items from. Whatever the case, I have found the best results came from having a purpose for why you are leveraging them as a mentor, and knowing what you hope to gain from the experience.

Later on, in my career, I started to realize the difference between a mentor and sponsor. For me personally, it wasn't until I was starting to position for the more senior-level positions that I realized the importance of sponsorship. Having someone in the decision-making seats who know you, the capabilities you bring to the table, the results you deliver, etc. and will actually vie for you is critical.

I now take great pride in mentoring and sponsoring both men and women. For one, I often learn more from the talented individuals I get the privilege to engage directly with. And as a female executive, being able to sponsor the next generation of leaders is one direct way I can give back and thank those who helped me get to where I am today.

Avoiding a Stall

How did you develop organizational savvy? And how did it help your career?

Developing organizational savviness is certainly not a skill we are born with. However, I do believe many of us are natural observers of our environments, and I think it is crucial we strive to understand what is going on both verbally and nonverbally. Throughout my career, I have always made it a point to really observe the environment I was operating in. For instance, what is the culture, who does most of the talking, who do people really listen to, what is the body language around the room during those times when others are talking, etc. When you take the time to really assess your environment, you can establish your role in that environment.

It is also an important skill for when you make a career move to really understand how to navigate the organization best to be able to deliver results. At the end of the day, we are all measured on the results we deliver, and now more than ever, we are also measured on how we deliver those results. By leveraging organizational savviness, you know the talent within the organization you want to have on your teams in order to be able to deliver the results. By understanding the culture and how you and your style fit within that culture, you can ensure you are delivering those results in a manner consistent with the values of your respective organization. I have been fortunate to be with a company whose corporate values align with my own personal values, so being able to show up as authentic and true to myself makes the "how" much easier.

Powering On

As you hit career obstacles, what motivated you to keep going?

Was there a significant event that changed your career trajectory and what was it?

Did you ever feel limited in a role and how did you circumvent that?

All of us have, at one time or another, felt we have been limited by a role. When I encountered that earlier in my career, I first made sure that the requirements of the job, the deliverables, and results I was responsible for were on time every time with the highest quality. I felt that as long as I was completing my assigned role, then opening the discussion for additional assignments and/or responsibilities was a much easier conversation.

For example, I was tapped once earlier in my career to take an opportunity to help a troubled program complete its certification process on a product that was severely behind both in cost and schedule. When I initially joined the team, the Director in charge of the program wanted me to focus on a very small subset of the overall remaining work to be completed. After several months on the effort, and observing the many areas of opportunity to streamline not only the remaining efforts but tie them together to improve the projected completion of the overall project, I sat down with the Director to have a candid discussion about my observations. Being able to show the results I had personally been delivering, their positive impact on the performance, and how I could connect those results and approach to possibly helping the overall project on a much larger scale allowed me to facilitate a "what-if" discussion.

My "what-if" was what if I took on these additional responsibilities to help him and the team be more successful. After a few discussions, he finally allowed me to move forward with my plan and gave me his full support. I was then able to capitalize on this opportunity to deliver results on a larger scale and put myself in a position where others within the company were taking notice of what I was doing. This would ultimately pave the way for me to take on my first Director role, overseeing my own product with full profit and loss responsibility.

I share this with you as an example of how you can take a role, maybe one that is not providing everything you are looking for, but you can look for ways to ask for additional responsibilities. The more you make yourself invaluable because of the results you deliver, the more you can turn the situation around for you and ultimately your career.

Annabel Flores

Vice President, Electronic Warfare Systems
Raytheon Intelligence and Space

About the Author

Annabel Flores is Vice President of Electronic Warfare Systems for Raytheon Intelligence and Space, a business of Raytheon Technologies. She oversees the overall strategic direction and operation of a diverse portfolio of electronic warfare products such as the United States (U.S.) Navy's Next-Generation Jammer and the U.S. Army's Electronic Warfare Planning and Management Tool, as well as the business's high-energy laser systems and Applied Signal Technology area.

Most recently, Flores served as the vice president of Electronic Warfare Systems at Raytheon Company's Space and Airborne Systems business prior to Raytheon Company's merger with United Technologies Corporation in 2020. She was previously Director of the Airborne Early Warning and Reconnaissance Systems product line within the Space and Airborne Systems Secure Sensor Solutions mission area. She has also been the manager for the Silent Knight Radar product line and held various business development roles at Raytheon Company's corporate team and with its Intelligence, Information, and Services business.

Annabel Flores

Flores was educated at the Massachusetts Institute of Technology in Cambridge, Massachusetts. She earned a dual Master's degree in Mechanical Engineering and an MBA through MIT's Leaders for Global Operations program. She also earned a Bachelor's degree in Mechanical Engineering from the institution.

Questions and Answers

Personal Career Insight

If you could give early-career aerospace women one piece of advice, what would it be?

Every career is filled with low and high points, and a support system is essential to getting through them and thriving.

For me, a small group of women who work in aerospace became my rock, a great soundboarding, and a sympathetic ear. As a group, our intent was to provide honest opinions and encouraging words to each other. Hearing about their career journeys and the challenges they faced, I gained perspective about my own path.

When we were all in the same city, we made it a point to meet for dinner. Regardless of the type of day or week I had, I knew we would have interesting conversations and return home inspired and recharged. When I crossed paths with them in the workplace, it helped to have a friendly face and ally in the room. Since we had previously talked about our personal growth and career goals, we could easily provide feedback to one another on everything from leadership style to team engagement.

Time and again, as we changed roles and moved to different cities, I relied on these women to get through the tough times and to celebrate the victories.

To any women starting their careers in this industry, I would encourage you to build your own support group, with like-minded people. This group can become your rock, helping you along on your career journey.

Work/Life Balance

Did you ever feel guilty for leaving work on time to attend a family event? Did you feel it reflected negatively on you at work? If so, how did you deal with that?

Like many people, I have struggled with balancing work and life throughout my career. On top of really enjoying the work that I do, I have workaholic tendencies—so there are times when it can be very difficult to unplug. Taking time for myself or to be with family can come with feelings of guilt, a voice-recording in my head nagging me that I've let someone down by not being at work.

What's worked for me over the years is learning to set boundaries and, because everyone is different, finding a balance that is right for me. If I set the expectation that I am available 24/7, people would grow to expect my constant presence. Instead, I have had to work at creating and, just as importantly, upholding those boundaries. The more I hold firm to my boundaries, the easier it is for me to set expectations with my team and leaders. It also lessens my own feelings of guilt if I leave work early or take time off.

As I continue my career, I carry a reminder with me—that every person struggles with work and life balance. I often share this reminder with my team, encouraging them to bring their whole selves to work and to be open about their priorities outside of work. I also tell them about my non-work priorities, hoping to make more genuine connections with people and to reinforce those respectful boundaries. I want my team to know that I fully understand the importance of family time, and that I support it, whether we have children or not. I believe this open dialogue helps encourage everyone to make the best decisions for themselves and their families.

Mentorship/Sponsorship

How important was mentorship/sponsorship for your career? Have they been men or women? How was the relationship established?

I was introduced to my very first mentor by my department manager at the time. This mentor, a program manager, was conscious of the fact that there weren't many women in our industry and wanted to support me. It helped that we had many things in common—we were both Latinas who grew up in small Texas towns, for example. Her willingness to help me and share her career journey and struggles set a great foundation for my own career. She solidified my desire to become a program manager and gave me the confidence I needed to strive for that goal.

After she retired, I created informal mentorships with leaders that I worked with and admired. As there are a limited number of female leaders in the industry, all of my subsequent mentors and sponsors have been men. Each mentor has helped me hone my leadership skills and has given me valuable guidance to map out my career. I often seek out people who have strengths in areas where I identify opportunities for growth. It helps guide the relationship with specific goals in mind.

Earlier in my career, I also joined my company's mentorship program, and I was assigned a senior executive as my sponsor. During those discussions, I shared specific gaps in my background and identified the types of experiences I needed in order to advance to executive levels. Laying this groundwork was key in helping my sponsor match me with the right roles in the right areas. All of my experiences with mentors have absolutely proven that mentors are invaluable for career success. From simple words of encouragement to sharing experiences, every bit counts.

Avoiding a Stall

Have you ever taken a role you were not excited about but had to show you were a "team player"? What was the outcome?

Early in my career, I had aspirations of becoming a program manager. When I finally took on the role, the opportunity quickly turned into a potential nightmare. As a deputy program manager, I was responsible for a new business opportunity, and my first assignment was to see a project through execution.

The stakes were high. We had to meet a critical customer need on an extremely challenging schedule—and on a fixed price. I was responsible for executing a program that statistically had no probability of success, and quite frankly, I was tempted to run away. But I recognized the importance of meeting the customer's need and decided to go all-in for my first shot at being a program manager.

To have any shot of meeting the customer commitment, I needed to convince the team—and myself—that this was a goal we could achieve. I engaged my entire team, as well as senior leaders, to ensure we had all of the resources needed to execute. The entire experience was a steep learning curve full of challenges to resolve and overcome.

In the end, we achieved our goal and met the customer commitment. A potential career derailment turned into one of my best professional experiences. It gave me the chance to build an extremely motivated, high-performing team, and it taught me the importance of focusing on the team's health in order to succeed. Finally, the role gave me the confidence to know that I can meet any challenge.

Powering On

Did you ever feel limited in a role and how did you circumvent that?

Over the course of my career, I have been in positions where I felt like I was no longer adding value to the role or growing professionally—my path on the corporate ladder unclear. As an engineer, I sometimes had a feeling that many big design decisions had already been made by the time I became involved. And I didn't always know how to approach those feelings of uncertainty, especially early in my career.

When an opportunity to transition to a new field arose, in Business Development (BD), I jumped at the chance. At first glance, BD would give me free rein in shaping solutions early and in designing products upfront. But it didn't take long for those feelings of uncertainty to return. I didn't feel as productive as when I was in engineering, and I was too far removed from delivering the end solution. I realized, after a process of self-reflection and guidance from a mentor, that to feel professionally fulfilled, I wanted to be in roles where I was constantly challenged and where I was accountable for those end solutions and results. So I returned to my earlier career dream of becoming a program manager. For me, it became the right balance of decision-making, technical content, and customer engagement.

Whenever feelings of uncertainty resurface, I engage in the same process of self-reflection, making sure I reach out to my mentors and support system. Through that lens, I am able to determine the root cause of the uncertainty and make career decisions with confidence.

Linda Flores

Nexus Program Manager
Bell

About the Author

Linda Flores serves as the Program Manager of Nexus at Bell. Linda and her team lay the groundwork for Bell's Nexus program by establishing the certification strategy, developing technology, and collaborating with community stakeholders on the infrastructure required for a safe, accessible, and sustainable mobility service.

Previously, Linda implemented the strategic alignment concept developed by General (RET.) Stanley McChrystal at Bell as it evolved its presence in the dynamic military and commercial vertical lift markets. Linda has also directed the continuous improvement and sustainment of the Bell 429 helicopter and led the airframe and integration design teams on the world's first fly-by-wire helicopter in the Bell 525. In addition to leading technical efforts in the commercial rotorcraft space, Linda has led military product development efforts at Lockheed Martin and Rockwell Collins.

Linda's career started at Lockheed Martin in the Leadership Development Program at their Missiles and Fire Control division. Linda held various positions in manufacturing, quality, and engineering where she was responsible for leading the systems engineering team for a multinational air defense program. At Rockwell Collins, she was responsible for leading systems engineering teams over fixed-wing, aircraft communication systems projects and unmanned vehicles.

Linda holds a Bachelor of Science in Industrial Engineering from Texas A&M University at Kingsville and a Master of Science in Systems Engineering from Southern Methodist University. Linda is passionate about serving her community where she leads scout dens and delivers meals as part of the Fort Worth Meals on Wheels program. She is an active mother of young boys that loves mountain biking, hiking, and traveling.

Questions and Answers

Personal Career Insight

If you could give early-career aerospace women one piece of advice, what would it be?

Don't be afraid. I work with some community youth groups in Texas, mentoring in science, technology, education, and math (STEM), and I see young girls who have an interest in a technical field but struggle to see themselves having a career in engineering. All too often I hear from girls that their familial circumstances will keep them from a successful career—whether they're a first-generation American or socio-economically disadvantaged. I share with them my background of growing up with a mother who grew up in foster homes and a father who spoke English as a second language and who fought every day to prove himself and make a life for his family. I'm not a product of the picture-perfect story they have in their heads, I am the product of working hard and not being afraid to go after what I want.

I put this philosophy to work every day still. Seeking out a career in aerospace, which has historically been a male-dominated industry, I knew it would not be an easy path. I knew I would be in the minority and I knew operational systems would likely be stacked against me. But after watching everything my father went through and overcame, I knew I had what it took to take things head on. For those already on the path in aerospace, my advice remains the same. Don't be afraid to really dig into it and commit yourself. There are perks to a technical career in aerospace. In my experience, there has been much more of a work/life balance and satisfaction in seeing my work come to life than if I would have chosen a more business-related role. It's also a major perk that your kids think you have the coolest job!

Work/Life Balance

Did you ever feel guilty for leaving work on time to attend a family event? Did you feel it reflected negatively on you at work? If so, how did you deal with that?

"Mom guilt" was never more apparent to me than when I had my first child. My son was born premature and had several health issues including a heart defect that made him more high-risk for Sudden Infant Death Syndrome (SIDS). He needed extra care and attention, but as the end of my maternity leave came closer, I found myself needing to choose between trusting a daycare to care for his fragile needs so I could return to work or quit my job to be with him until he grew stronger. It was an impossible decision to make, but in the end, I submitted my letter of resignation and was open and honest with my leadership team on my personal struggles.

Much to my surprise, my honesty and transparency was met with support from the organization. They vowed to find a way to make it work and shortly after a corporate work-from-home policy was rolled out that allowed me to work remotely during those crucial months with my son. I think there is often a lot of fear and uncertainty with women who are often pulling double (or triple) duty caring for their children, their parents, or other family members and the perception of being a caretaker has on their career. As an introvert, it was difficult to come forward. But what I learned is that I owe it to myself, and to the other women facing similar dilemmas, to bring these issues into the light and be a part of a cultural shift. Most importantly, no matter what season of your life you're going through, know there is support out there—you just have to ask.

Mentorship/Sponsorship

If you had a Sponsor to reach executive levels, how did you meet that person? Was the person assigned to you or was it a boss that became your sponsor? How did this person help your trajectory?

I had three mentors while at Bell—two were male and one was female—that were assigned to me through the Bell mentorship program. I really won the lottery as two of those mentors went on to become incredible aerospace leaders.

My first mentor was Tom Wood. When we met, he was much older and was a long-standing technical expert at Bell. There was nothing Tom didn't know about Bell's programs. He seemed entertained by mentoring me, whether that was because I was a woman or because we both kept each other on our toes. Tom laid a very important foundation for me, as I was new to aviation at the time. He really taught me to seek out answers to the unknown. If I didn't understand a process or a concept, he would help point me in a direction but encourage me to take the initiative to find the answers and connect the dots for myself. Tom has since retired from Bell, but his lessons and wisdom he instilled early in my career continue to stay with me.

My next mentor was Mitch Snyder, who at the time was the Executive Vice President of Military Business. He was, and still is, a proponent for more women in leadership roles at Bell. His mentoring style was much different than what I was accustomed to as he spent more time asking questions than giving advice. He took time to understand what it was like to be in my shoes and I was able to offer insight on how to attract other females to this industry and the trials and tribulations women, especially new moms, have in the workplace. He showed me how powerful it could be to to show compassion and lead with empathy; things females are usually afraid to do because of the risk that you may be perceived as weak. He was always inquisitive and a champion for making sure everyone feels welcome at the company.

Mitch is now the President and CEO of Bell. My last mentor was Lisa Atherton, who became the Executive Vice President of Military Business when Mitch became CEO. She was an amazing sounding board and an excellent example for women in this industry to look up to—someone who was a great example of work/life balance between her career and family. She was someone I could be open and transparent with, as a model of female strength and confidence in a male-dominated industry. She made sure I walked away knowing that the personal pressure we set on ourselves to be perfect is what often leads to insecurities that diminish our strength. Lisa is now the President and CEO of Textron Systems, a sister company of Bell.

Each of these mentors taught me something different and shaped my trajectory throughout my career—from getting my bearings in a new industry to navigating a new career as a woman in aerospace and becoming a voice and advocate for other women around me.

Avoiding a Stall

Have you ever taken a role you were not excited about but had to show you were a "team player"? What was the outcome?

I'll be honest, when I came to Bell it was for a position that I wasn't completely excited about. In my former role at another company, I was commuting an hour each way to work. I wanted to find more balance in my life and spend more time with my family, so I took a map and drew a small radius around my home and began looking for opportunities within it. I landed a job at Bell in configuration management; however, the role wasn't very technical, and for someone with a background in engineering who was formerly leading a missiles program, it also wasn't very exciting. I spent a year in the role and absorbed and learned everything I could about Bell and its programs. I eventually took this as an opportunity to begin carving my own path.

At the time SAP was new to the organization and I began playing with custom engineering dashboards. It was a new tool that wasn't being utilized, so I put my skillset of combing data to work on top of what I was doing as part of my "job." Soon after I received a call from the CEO requesting specialized dashboards showing engineering performance. I quickly found out that I was fulfilling a need the organization didn't know they had. Working with leadership this way allowed for conversations about what I did before coming to Bell and where I saw my career going. By going above and beyond and finding additional ways to be useful and a team player was a way for me to be seen and opened the door to other possibilities.

Powering On

Did you ever feel limited in a role and how did you circumvent that?

Before joining Bell, my background had been in missile development and aircraft communication for air defense. Coming into the rotorcraft world, I felt out of my depth—I had no background in helicopter dynamics at all, and I joined a team that was incredibly bright. I had a lot of fear that the entire team had a lot more background and knowledge than me and that there wasn't much that I could bring to the table. However, it was this "impostor syndrome" that kept me eager and hungry to learn more and quickly get up to speed.

Although I was familiar with a lot of the principles, the industry was completely new to me. I knew that I would need to reeducate and reinvent myself for this new realm of aerospace. I started by building a support system. I used my mentor at the time to identify where I could seek out answers. It took a lot of effort but I was committed. I quickly began to build relationships that would guide me and create a tribe of people that I could turn to with questions or for advice, and it paid off. Having a group of people in your corner, who are supportive, is critical.

Mary Lee Gambone

Head of Materials Engineering
Rolls-Royce

About the Author

D r. Mary Lee Gambone is currently Rolls-Royce Head of Materials Engineering for their Defense business, leading a team of more than 100 materials engineers in both the US and the UK.

Mary Lee has worked in aerospace materials for more than 35 years as both a technical specialist and engineering leader, developing and implementing new materials and manufacturing techniques to enable higher performance jet engines. Mary Lee earned a BS in Metallurgical Engineering from Purdue University in 1982, where she also was a co-op for Armco Steel. In 1984 she earned an MS in Materials Engineering from Massachusetts Institute of Technology, where she completed a thesis focused on nitrogen strengthening in austenitic stainless steel.

Her first full-time professional role was with Allison Gas Turbine Division of General Motors Corporation. She focused on mechanical testing, primarily fatigue, and fracture mechanics to create design data for metals and contributed to material development, including advanced nickel-based alloys and titanium intermetallic matrix composites for gas turbine engine applications. She left Allison in 1988, and, after a brief tenure at a small materials-testing company, she earned a PhD in Materials Science at the University of Virginia (UVA).

While at UVA, she joined the US Air Force Research Laboratory (AFRL) at Wright-Patterson Air Force Base, who supported her dissertation research on the impact of SiC fiber strength distribution on creep of titanium matrix composites. After finishing her degree, she served as Team Lead and Research Scientist focused on understanding the properties of metal matrix composites in the Metals Development

Branch of AFRL. She left AFRL and joined Rolls-Royce in 1998 to work in material and manufacturing technology development. At Rolls-Royce, she has held several roles, including Manager of Critical Part Lifing and Chief of Research and Technology Strategy. She was chosen to head materials engineering in Indianapolis in 2010. In this role she has also served on the Executive Steering Committee of the AFRL sponsored Metal Affordability Initiative (MAI) Consortium and as President of the industry-only MAI Association.

She has coordinated the Rolls-Royce University Technology Partnership with Purdue University and served on the industry advisory committee for the Purdue University's School of Materials Science and Engineering. She is also an advisor to the Rolls-Royce Women North America employee resource group and a facilitator and master champion for High-Performance Cultural engagement for the corporation.

Questions and Answers

Personal Career Insight

How did you decide between a leadership and technical career track?

I think it is important to be open to exploring different "tracks" in one's career. I have had great work experiences and fulfilling roles both in technical specialism and in people leadership. I have learned valuable life lessons and gained insights about myself in both types of jobs. I didn't begin my career thinking, "I am going to be a research scientist" or "I am going to be Head of Engineering." In my early career, I didn't really think anything about my future beyond wanting to do a good job of solving the problems in front of me at the time. I was excited by challenging technical problems and stimulated by the engineering colleagues I worked with, and that was enough.

In my early thirties, I completed a PhD dissertation. That research didn't go as I expected it to, but some really interesting questions came out of the results. I wanted to work on some of those questions, and then life happened—I had other job responsibilities and a new baby—those things supplanted the time to think about fundamental questions. Not to say I wasn't happy in my career. I was doing interesting work. What I found, though, was a real passion for engaging with other scientists and engineers, listening to their ideas, and helping them figure out how to test them and resolve issues.

I had one colleague who struggled with presenting too much detail in technical reviews. He would lose his audience when he overwhelmed them with data. I coached him to pull out his main messages and present the data that best underpinned those messages. I had another colleague that struggled to finish his projects. He would start with great ideas and then get bogged down in the middle. Together we created a series of mini goals that helped him stay on a course to complete his larger projects.

From these and other opportunities to help my co-workers be more successful, I learned that I am energized by helping others and that, by doing so, my impact is amplified beyond the technical work that I alone can do. Knowing this about myself, as I progressed through my career, I have chosen opportunities to lead engineering teams over technical specialty. I can say that today I am in a role that is fulfilling and makes me happy.

Work/Life Balance

Did you ever have to make a move to advance your career that impacted your family life and how did you balance the two?

The most significant threat to the stability of my marriage grew out of the advancements of both my husband's and my own career. We were both promoted to management roles within months of our second daughter being born. We had a six-year-old, a newborn, and dramatically increased responsibilities at work all at the same time. We had a plan: We hired a nanny who could take our older daughter to school and care for our infant in our home.

Even the best-laid plans can fall apart. I had to do some travel as part of my new role, and my husband was on call nearly 24/7, as he was managing a production department. The job pressure for both of us was significantly higher than it had been. It also turned out that the nanny was not very competent. The biggest problem, though, was that neither of us raised a hand and said, "This isn't working," until there was a complete crisis. One week while I was traveling, our older daughter became ill. My husband had to leave work to take her to the doctor, and when he got home, he found the whole house in chaos because the nanny wasn't doing her job. He called me that night at the hotel and told me I had to get home or else. I cut my trip short and flew home. He was angry; I was angry. We each felt like we were the victim and the other was being unreasonable.

Mary Lee Gambone

Luckily, we pulled back from the brink. My husband had a mentor at work that he talked to, and I agreed to not travel for the time being. We fired the nanny and found another one that was highly competent. I understand now that I was demonstrating plan continuation bias—even though our carefully constructed work-life balance plan was falling apart, I didn't want to change course or pause to rework the plan. I didn't heed the signals that my husband was giving me that the plan wasn't working for him. The way we got to balance was to communicate more fully and more often. When either of us had something big going on at work, we would talk it through, so neither was left with too much to handle. I also accepted that I would be the primary parent that interfaced with caregivers and school. My husband did more around the house in the evenings and weekend to compensate—laundry, cooking, etc. This worked for the last 18 years, and now we can look forward to the challenges of retirement in the next few years!

Mentorship/Sponsorship

What activities have you engaged in that have helped other women achieve success in their aerospace careers?

I think the most important way any leader in aerospace, or any industry for that matter, can help women achieve success is to see them and work with them as the professionals that they are, rather than as women professionals. When I think of the frustrating periods in my own career when I felt constrained because I was a woman, they are all defined by someone in my leadership who, instead of assessing me for my knowledge and experience, saw me as a woman first and let that drive their decisions. To combat this behavior I have tried to set an example for my male colleagues, and I have become increasingly more vocal to senior leadership when I see systematic gender bias.

To set an example I strive to interview, hire, and promote to create the most gender-diverse organization possible. This is not difficult, as talented and capable women abound. I have never had to select a candidate for a role that was less qualified in order to create gender diversity. Indeed, I often find myself having to select between two exceptionally qualified women candidates! I also mentor women engineers at all career stages to offer advice, but even more than advice, encouragement to take risks and to seek new roles or promotions. I try to coach them to be more positively self-promoting and confident in their capabilities. I never received this kind of coaching in my early career, and I think self-confidence is crucial to breaking down gender barriers.

I also find myself more often challenging long-standing assumptions about how succession plans and candidate lists are created. In engineering, particularly for the highest-level roles, these lists are very much based on who the incumbents "know," not who is most qualified for the role. It can be a bit disheartening to be the lone voice in the room calling out a woman's name who has been overlooked. Even so, it is vitally important to keep doing it! I used to worry that my male colleagues would see me as biased towards women. I have let that fear go as I get closer to retirement. The only way we will ever achieve any kind of gender equity in aerospace is to get qualified and deserving women into consideration for leadership roles. Once they are on the list, I am confident they will earn those roles, and my lone voice will be joined by theirs!

Avoiding a Stall

How did you develop organizational savvy? And how did it help your career?

I don't think I ever consciously set out to develop organizational savvy. When you have a long and successful career, though, you do. When I was a young engineer, I remember listening to a series of lectures on cassette tapes, called "Managing Conflict Productively." One of the main points the speaker made was that an individual has power in a relationship equal to the value of whatever they provide to the other person in that relationship. Seen this way, even the lowest-level employee may have power in their relationship with their manager, if they provide something that their manager cannot get from anyone else.

This was a revelation for me because at the time I could not get any attention or consideration from my manager. When I thought about the situation from the perspective of power, I realized that my manager was not interested or engaged in the technical work I was doing—he cared deeply about other work in the group that he had started before being promoted to leadership, but my area of focus was of no interest to him. This was a lightning bolt of insight, and I made plans from that moment to find another role, which was the right move for me at the time.

I have since used this principle to understand better my relationships with my colleagues, leaders, and staff, as well as customers and suppliers. When I am frustrated with how a project is going, I step back and try to think about what the different parties involved are wanting from the project and what they define as success. I find that thinking about problems from this perspective often leads me to alternative and more productive approaches to overcoming them.

Powering On

Did you ever feel limited in a role and how did you circumvent that?

Until I earned the Head of Materials Engineering role I have today, I have had jobs that grew to be limiting or, even worse, boring. For the first five years in my first engineering role, I was engaged and excited. I got opportunities to grow my skills, understanding, and responsibility. Then there was a reorganization, and I was moved to a new manager who seemed to have little interest in my work or development. So I left that company and tried something new. I have left jobs and organizations several other times in my career, usually when I felt stymied and saw no way to progress.

Of course, once I was married and had children, quitting a job and starting fresh was not as viable an option. Instead, I have made an effort to move within the company I worked for rather than leave it. I have also had to develop alternative approaches to gain more job satisfaction. If I find myself unhappy in a role, I think about why I am unhappy. I have had jobs that were not challenging and ones that I felt didn't fit my skills. In those circumstances, I have tried to supplement my specific objectives with activities, which might benefit the company as well as allowing me to grow. Sometimes this has meant stretching myself technically by writing papers or presenting at symposia. Other times I have taken on diverse challenges such as acting as a coordinator for university research, serving on the leadership committee of one of our Diversity and Inclusion groups, and facilitating high-performance culture engagement sessions.

Through these sorts of activities, I have built my corporate and professional network and demonstrated that I have leadership capability that may not have been apparent from the strict description of my current job. In this way, when a new, more exciting opportunity presented itself, I felt prepared to apply and make my case as the best candidate.

Neri Ganzarski

Student, Mechanical Engineering
Washington State University

About the Author

Neri Ganzarski is a student at Washington State University (WSU) majoring in Mechanical Engineering with a concentration in Autonomous Systems. She is a research assistant helping with the testing of a steerable medical robot, a member of WSU's Formula SAE race team working on composites and aerodynamics, and a peer mentor in the "Intro to Engineering" class. Neri is a WSU "Top Scholar" and a President's Honor Roll student.

Neri Ganzarski

Neri found her love for engineering after building her first robot in a high school class. Together with her friend (and the only other girl in the engineering class), she built a simple delivery robot to deliver books between classrooms. It was then that Neri knew robotics was the place for her.

Neri was a competitive swimmer for eight years before a significant back injury cut that passion short. But through perseverance, she found herself in the pool again, this time as a swim instructor teaching kids and adults correct swimming techniques to avoid injuries like hers.

Neri's family is originally from Israel, and she loves traveling there to visit extended family. She loves spending free time with her mom (Einat), dad (Roei), two younger siblings (Omri and Daniel), and their dogs, goats, chickens, and rabbits.

Questions and Answers

Personal Career Insight

What was the coolest thing you experienced in your career?

I got the opportunity to work on a summer project for an electric aviation propulsion company. At the time, they were very busy in their final stages of design and test, working nonstop to build and integrate their technology. In addition, they were interacting with customers and regulatory authorities and needed high-quality documentation to share. But they had no time to create any detailed external-facing renders and needed assistance to make important deadlines in delivering those external facing documents.

I used what I learned at WSU to create quality renders for their motors, as well as design some small add-on parts such as oil opening covers that were needed to finish the models.

This was an exciting project for me because it was at WSU not just a class assignment to complete, rather it was something that a real company was actually going to use for their business. I had to make sure that everything was up to their highest standard.

By midsummer, my final renders were incorporated into the company's external-facing document, and were sent out to their customers and regulatory authorities.

Doing this summer project was the coolest thing I've ever experienced related to engineering. And as an added bonus, I got to witness the first flight of the world's largest all-electric commercial aircraft.

Work/Life Balance

Did you ever have to make a move to advance your career (within your company or changing companies) that impacted your family life, and how did you balance the two?

One of the lessons my parents always taught me was to always do the right thing even when it's hard. The process of choosing which university I'd attend to pursue my passion for robotics and get my degree in Mechanical Engineering was one of the hardest decisions I've ever had to make.

At the time, I was accepted into a university close to home, but not to their engineering program. I was also accepted into other universities directly into their engineering programs. The downside was they were far away from home and my family.

Neri Ganzarski

Getting into the engineering school at the university next to home would have allowed me to "dance at two weddings" at once. But that was not to be and I was faced with my toughest decision to make—I had to choose between attending a university close to home (but not study engineering) or attending a university that was far away from home, but study what I love.

Now, for some of you readers, this might seem like a silly dilemma. Childish even. But as an 18-year-old girl who is very close to her family, this was a very hard decision to make. It would completely change my family life. If I pursued my passion, I would no longer be part of our daily family dinners where we share what we did that day, I wouldn't be part of movie night where we spent half the time arguing what movie we would watch together, and I wouldn't be able to go to them if I simply missed them or needed some advice. Pursing my passion was not a simple decision.

I weighed the pros and cons of each option and eventually decided to do what seemed at the time, as an amazingly hard thing—leave home and move to a faraway campus to pursue engineering. I am now studying mechanical engineering and having the time of my life. Between Formula SAE race-car building and being a teaching assistant in the Intro to Engineering class, I keep quite busy and am I learning a lot.

But what about my family? At first, I thought I'd be too busy with school, and not have enough time to talk to my family, which obviously at the time, made the thought of going there so much harder. But as I got more comfortable with my routine and schedule, I ended up finding a lot of time to talk with my siblings and parents. Whether it was on my way to classes, every night before I went to sleep, or while I was making food, I ended up finding time to be both an engineering student and keep in close touch with my family and not lose that close connection.

In life, we don't always get everything we want. Sometimes we find ourselves in a position of having to make that tough choice between options, each of which is not perfect. Weigh the pros and cons of each, and commit fully and wholly to the decision you end up making. You will find, as I have, that committing to something, whatever it may be, brings you satisfaction, learning, and fulfillment.

Mentorship/Sponsorship

What activities have you engaged in that have helped other women achieve success in their aerospace careers?

I love robotics. It's my passion. In my 11th grade of high school, I wanted to join a robotics club. But in a club filled with boy's teams, the girls were clearly and sadly not present. I decided to form an all-girls robotics team so that I and other girls my age could have the opportunity to pursue their passion in engineering and robotics.

I had to find girls who were passionate about engineering (or didn't yet know that they were) and were willing to join a team. This was much easier said than done. Little did I know that a lot of the girls I talked to experienced various challenges in the engineering field and were left discouraged from following their passion. Like me, they felt that a girl succeeding in engineering was going to be an uphill battle.

I had to explain to them that the reason I was trying to create an all-girls robotics team was to give them an opportunity to follow their dreams without the fear of challenges or hurdles just for being girls. We would have a lot of technical issues to overcome, but it wouldn't be due to our gender. We ended up creating the team and working on robots for the rest of the year.

As I transitioned from high school to university, I wanted to continue helping other young women feel motivated, included, and valued in an industry that, at times, does the opposite.

I joined my university's Formula SAE race team with my friend, only to find that we were the only two engineering girls in a group of fifty boys. It was intimidating at first but the boys treated me like any other engineering student. I wanted to let other girls know that car racing was not "a boy's thing" and that building cars was a great way to practice engineering skills. And that perhaps if more girls joined, it would not be such a male-dominated club. Whenever I meet new female engineering students around campus, I invite them to our team meetings, show them our race car and the workshop, and encourage them to join the team. I also interviewed for the university newspaper representing our racing team so that other girls might be encouraged to get involved in such activities.

In a field dominated by men, one way I have found to help other young women is to lead by example and take action. And while I hope this will no longer be required in the future, I plan to continue doing so as long as it is required.

Avoiding a Stall

Have you ever taken a role you were not excited about but had to show you were a "team player"? What was the outcome?

Throughout school, work, life, and, in my case, being an engineering student, a majority of assignments and projects are team based. Bouncing ideas off of group members, asking thought-provoking questions, seeing things through different perspectives, all while learning how to work with diverse groups of people, is what makes us learn and grow, no matter where we might be.

With that said, there are times, when we need to work with people that might not always make things easy—people or groups, who might not be as excited or devoted to the assignment or task as you are. But that is just part of life. Learning to work with everyone and anyone, finding new ways to conquer problems, no matter how exciting, or inconvenient things might be is what being a team player means. It is not about me, or you, but rather about the team.

I remember being on a team that was tasked with designing a mechanical device. My team members pushed for a design that I did not agree with or thought would work. But since the rest of my team was excited about their design, I agreed to it as well. We all worked together on the design and on building the device. And while it only partially worked, we learned a lot as a team. That in itself was worth the effort.

Powering On

As you hit career obstacles, what motivated you to keep going?

A couple years ago, I joined a competitive robotics team. The first month of being on the team was amazing. It was invigorating, challenging, and fun, all at the same time. I helped build autonomous and remote-controlled robots.

But despite having a great time there was always something that didn't seem right. There was always a sensation that left me feeling excluded from the rest of the group, and I wasn't sure why. By the time the second month rolled around, I figured it out.

One day I received a shocking text from my team captain asking me to leave the team because as the only girl on an all-boys team I apparently caused "too much of a distraction" for the boys.

I wasn't sure what to think. Did I do something wrong? Could being a girl in this day and age really be that much of a problem for boys in a robotics club? Apparently yes!

I loved robotics too much to quit. I wanted to prove to those boys and to myself that I was more than the "distraction" they saw me as. This is why I decided to start my own team, an all-girls robotics team that would include girls like me who have struggled fitting in throughout this field.

I worked tirelessly for months to recruit girls and create a team. It was not easy. I had to push through many obstacles, including ridiculous ones, like my own high school administrator (a woman by the way) suggesting that I should look at doing something easier than creating my own robotics team. But I wanted to do robotics, and no one had the right to stop me.

As the obstacles kept coming, my perseverance grew stronger, and I eventually succeeded in creating an all-girls team. This whole story came full circle when an all-boys team from that previous club asked if they could merge teams with mine and have me be the captain of the combined co-ed team—I agreed.

I learned a lot from this experience. I learned to never let anyone decide what I can and can't do. I learned to welcome everybody and anybody who wanted to be part of something they were passionate about, no matter who they were. I learned to never give up, even if it was hard, and seemed like nothing would work. And I learned to push through challenging obstacles, work hard for what I want and fight hard for what I believe in. These are lessons I will never forget.

Joan Higginbotham

Retired Astronaut
NASA

About the Author

The Chicago native didn't set out to use her newly honed electrical engineering skills in the space industry. But that changed the minute she saw a space shuttle and the launch pads, which looked like something out of a *Star Wars* episode. Just two weeks after graduating college, Joan began her career as a rocket scientist at the NASA Kennedy Space Center in Florida, working on the shuttles' electrical systems. She went on to hold a number of positions, including working in the firing room—the launch control "nerve center" during the shuttle countdown and liftoff. Always thirsting for knowledge, she earned a Master of Science degree in Engineering Management from Florida Institute of Technology. At the urging of her boss, she applied to join the Astronaut Corps in 1994. Not being successful, she decided to earn another degree, this time in space systems. She reapplied and was selected to the 1996 Astronaut Candidate class.

During her nine-year tenure at Kennedy, she actively participated in 53 space shuttle launches—an impressive accomplishment for anyone. However, when she returned to Kennedy for the launch of Discovery on STS-116, she took "participation" to a whole new level, as astronaut Joan Higginbotham. On her nearly 13-day space mission, Joan's primary task was to operate the Space Station Remote Manipulator System, better known as the robotic arm, assisting with the installation of the P5 truss and supporting crew members during the spacewalks to rewire the space station's power system and retract a solar panel. To date, she is the third of only three African-American women astronauts to fly in space.

Higginbotham has received numerous awards over her 30-plus-year career including World Who's Who of Women, Charlotte's (NC) 50 Most Influential Women, the National Space Medal, the NASA Exceptional Service Medal, and Savoy Magazine's 2012 Top Influential Women in Corporate America. She's also been honored with an Honorary Doctorate in Aerospace Science from her alma mater Southern Illinois University at Carbondale and an Honorary Doctorate in Humanities from the University of New Orleans. She resides in Charlotte, NC, and is married to the Honorable James E. Mitchell, Jr.

Questions and Answers

Personal Career Insight

What was the coolest thing you experienced in your career?

The coolest thing I've experienced was flying in space for 13 days; but not for the most obvious reason of "it's just plain cool".

Our space shuttle mission was extremely busy for the first seven days, and every crew member's schedule was jam-packed. It wasn't until the 8th day that we had some free time, so I floated over to the Russian segment, pressed my face against the window, and watched the world go by at 17,500 mph.

As we were flying over the Middle East, I saw something that looked like smoke billowing up from the ground. I immediately thought there was some unrest...perhaps one entity was bombing another. Then "it" hit me like a ton of bricks! I turned my head and looked back inside the International Space Station and thought, "Here we are, this mini 'united nations' crew composed of two African American astronauts, a Scandinavian astronaut, an astronaut of Indian descent, an astronaut of Jewish and Korean descent, a German astronaut, a Spanish-American astronaut, an astronaut born in the UK who became an American citizen in pursuit of his dream of becoming an astronaut, and a Russian cosmonaut." We spoke five languages, represented four countries and three space agencies. I thought, "If the ten of us could get along and work together for a common good while confined in this tin can of a spaceship, why can't we all get along here on Earth, where there is so much more space?" It was quite a profound moment for me.

Work/Life Balance

Did you ever feel guilty for leaving work on time to attend a family event? Did you feel it reflected negatively on you at work? If so, how did you deal with that?

Here is my opinion of work/life balance. It's a fallacy, at least for me. At any given time, there is one thing in my life that takes precedence over another thing; and the pendulum swings back and forth. But I believe it all evens out in the end.

Even in 2020, I believe that principally, women remain the primary caretakers of their families. Many of those women like myself have careers, and we precariously balance ourselves on the work/life scale. And we will go to great lengths to make everything work.

Joan Higginbotham

Me and my husband, James, supporting the Carolina
Panther's football team.

I had just transitioned to a new role, and the team and I were in the planning stages of hosting a day-long on-campus "shark tank-esque" event for small and diverse vendors, which was a phenomenal opportunity for those businesses. At the same time, my stepdaughter was killing it at her speech and debate competitions, so much so, that in her first year of competing, she qualified for the National Competition in Utah. As life would have it, the week of nationals was the same week as our vendor event. There was no way I could miss the vendor event and there was no way I could miss nationals.

Joan Higginbotham

Me repelling off a cliff in the Canyonlands of Moab, Utah.

I flew out to Utah and was there to provide support on the first two days of competition. I caught the red-eye out of Utah that second night, landed the next morning, and drove straight from the airport to work to finish overseeing the setup for our vendor event scheduled for the following day. Was I tired as all get out? Yes, but I accomplished both goals of supporting my stepdaughter and fulfilling my work obligation. As I said, we will go to great lengths to make everything work.

There are times I must leave work early, and it's understood by my colleagues and my manager that either I came in early, or I'll hop online later, but they know my work will get done…on time. They know and trust that I am a reliable, competent worker, and regardless of what's going on in my personal life, the job will get done, as flawlessly as humanly possible. Therefore, they don't worry when I leave work early or on time. My advice is to work for a company like that.

Mentorship/Sponsorship

How important was mentorship/sponsorship for your career? Have they been men or women? How was the relationship established?

There are several types of people that are important to me in my career: A *mentor* whose career I'd like to emulate and helps develop my career. A *coach* who provides specific feedback designed to improve my job performance. The coaching role is more comprehensive and time intensive than the mentoring role. Finally, I have an *advocate*, an executive in the organization who speaks on my behalf when I'm not in the room, and in rooms where I normally wouldn't be (such as an executive leadership team meeting). I encourage people, especially young women who are early in their career, to establish one or more of these relationships as they are incredibly beneficial and, in my mind, essential as you progress through your career.

As I reflect back, most of these roles have been primarily filled by men, partly due to the fact that I've worked in male-dominated fields for the majority of my career. However, I have had two women who have provided tremendous support to me during my career.

These invaluable relationships I've had with mentors, coaches, and advocates have formed both organically, where the individual and I just "clicked" and through intentionality, when I've asked a particular individual if he/she would serve in a specific capacity. Regardless of how the relationships have formed, I have been incredibly blessed to have these people in my life.

Avoiding a Stall

Have you ever taken a role you were not excited about but had to show you were a "team player"? What was the outcome?

I was working for a company that went through a major company-wide organizational shake-up. The result of the reorg was two-thirds of my team members were involuntarily separated from the company. I found out about this in a way that no one ever should. My (former) team member called to say that "it was nice working with the team." Several hundred employees were walked out of the door that day. Needless to say, this sent shockwaves throughout the work campus and elicited a slew of emotions on my part, none of them very positive. The following day, I was assigned three new areas of responsibility with three brand new teams.

Although the initial shock of the whole ordeal began to wane, there were lingering feelings of confusion, uncertainty, and a little despair, even though we were among the lucky ones still standing. The team wanted some assurance that the bloodletting was complete, and their jobs were secure. I couldn't provide that reassurance, but I did promise to be transparent with them.

My first course of action was to go about learning the new businesses and begin building one unified team from three disparate teams. It was challenging and stressful, to say the least. We were on the long journey to achieving the vision for our team when an opportunity too good to pass up came my way. I ultimately left the company to pursue this amazing opportunity.

Powering On

As you hit career obstacles, what motivated you to keep going?

Life is fraught with obstacles, both on a personal and professional level, so the first thing I've done is to acknowledge and accept that. Whenever there has been a hiccup in my career, I would take a step back and first get a handle on my emotional state, because that directly impacted how I approached the issue. Once I'm in a place where I can be objective, I put on my engineer's hat and utilize my analytical prowess to approach the obstacle as a problem that needs to be solved. Then I choose to view it as an opportunity to grow. I develop a strategy to address the obstacle or devise a way to go around it, execute that strategy, and then assess how well it is working. It's an iterative process that requires me to continuously reassess the strategy and its effectiveness. That's the practical way I approach the problem.

The second way I contend with obstacles is through faith. I am a Christian and stay true to what I've been taught all of my life, and that is to turn to God. So I have a little talk with God, trust that He is in control, and know that at the end of the day "this too shall pass." No obstacle is insurmountable.

Raelyn Holmes

Global Trade Compliance Manager
Pratt & Whitney

About the Author

Raelyn Holmes is currently a Global Trade Compliance Manager in the aerospace industry. She began her career in aerospace after obtaining a BS in Mechanical Engineering from Tuskegee University. She is a native Texan residing in South Florida.

She has demonstrated specific strengths in the areas of project planning/execution, event facilitation, continuous improvement, international trade, and customer relations over a 20-year continuous service with United Technologies Corporation, now Raytheon Technologies Corporation (RTX).

Over the years, she has taken on increasing levels of responsibility in programs including Space Shuttle Main Engines, Hypersonic and Advanced Programs, and International Auxiliary Power Systems. Her roles have included Program Manager, Project Engineer, Continuous Improvement Manager (and Facilitator), and International Trade Compliance Manager.

She has dedicated herself to support and lead initiatives that improve work environments, employee morale and diversity, mission safety, and customer satisfaction.

Continuing her education while working, she obtained an MS in Quality Management Systems and MS in Technical Management at the National Graduate School and Embry-Riddle Aeronautical University, respectively. She also studied in the master's programs for Mechanical and Industrial Engineering at the University of Miami.

She is a mother, as well as, an active life member of Alpha Kappa Alpha Sorority, Inc., Tuskegee National Alumni Association, various other community initiatives, and President of the African-American Forum (an RTX company employee resource group focused on influencing diversity of thought).

Her hobbies include restoring a 1981 CJ7 Jeep, supporting her Girl Scout's Sporting Events and School PTA, celebrating with family/friends, collecting music albums/stamps, attending plays/musicals/concerts, golfing, fishing, and singing in the choir. Raelyn's favorite games are Monopoly, Toy Blast (app), dominoes, various card games, Atari's Pac-Man/Miss Pac-Man, and chess.

Questions and Answers

Personal Career Insight

What inspired you to choose a career in Aerospace and what has given you the most satisfaction during your career?

As a child, I played, tinkered, created, and explored my section of Texas. I also found that the math and science subjects resonated with me, and when I had challenges, my parents or their friends were able to help me. When I started to think of a career and field of study, I listened to and watched my family, which ultimately led me to choose a career in engineering; aerospace was a focus when I considered job opportunities.

I recall my aunt reminding me to select a career that could "keep me in the lifestyle I had become accustomed to" (spoken in her southern dialect). My family was all abuzz with news about my cousin's success; he was the General Manager of Austin Energy. My neighbor worked for NASA as a Flight Design Integrator and Manager, and was a working mother/wife who supported me throughout my life. Upon finishing my Mechanical Engineering degree, I accepted the opportunity to work on NASA's Space Shuttle Main Engine (SSME) Program through Pratt & Whitney's Liquid Propulsion business; it was the type of meaningful work that I needed. On my first day with Pratt & Whitney, the team was supporting the launch of STS-85. As I watched the launch from the Pratt & Whitney SSME High-Pressure Turbopump Control Room, I remember the emotion that overcame me as knew that I was now a part of something bigger than me, my work would be valued, and lives depended on me doing my job well. I also appreciated that as a little girl growing up in Houston, I had never imagined working on the Space Shuttle, and now the technology that was in my backyard and television screens was now on my desk.

I think the most satisfaction in my Aerospace career has come from volunteering and recruitment activities, such as yearly summer Rocketry Competitions at Tuskegee University's FASTREC Program, hopefully resulting in meaningful impacts for the students, administrators, and the diversity of our future workforce and technological advancements. During those years of competitions and various activities with the university, our team not only sacrificed time from their families but did something new each year with the students. The competition expectations, including technical, team dynamics, schedule expectations, increased with each offering. We were not always sure that "we" or the students could meet the demands; however, each year we all surprised ourselves with the successes and exposure! In general, the advice is: do not be intimidated to try something new or unusual, which applies in our volunteer efforts as well as our professional and personal endeavors.

Work/Life Balance

Did you ever have to make a move to advance your career (within your company or changing companies) that impacted your family life, and how did you balance the two?

I have moved within the company to advance my career. Each move has been unique and encompassed different challenges; some dynamics are too numerous or painful to remember. In 2009, I was a Program Manager for an Advanced Hypersonic Programs in West Palm Beach, FL. It was a challenging and rewarding period of time. An opportunity to lead continuous improvement at a sister company in San Diego, California arose; it was an opportunity to work on one of my goals and live in another desirable area. I was offered the job, and while preparing for the move, I discovered that I was expecting the birth of my daughter (my first child). I was 35 years old, now a single woman, and excited about motherhood.

Raelyn Holmes

Daughter, Ava VanZandt Holmes, and our dog, Jersey, pose for picture (2015).

The move proved to be challenging, but my family and friends' support was substantial. My mother and cousin moved from Texas to Florida, and finally to California to help me pack and settle into motherhood. The move also required some special consideration from the company as well. My medical condition worsened during the moving period, schedules were forced to adjust, and extended duration in temporary housing was required. I settled into the new city, company, and working environment. During my time there I was offered a Program Manager position for an international commercial aircraft program; it was a position that I ultimately declined despite the opportunity being in line with my career goals and my love of traveling. The position would have required more of my time and personal sacrifice than I, my daughter, and family, could afford.

Ava and I at Tuskegee University with the
Booker T. Washington Monument (2014)

Raelyn Holmes

I received feedback from my mentor that declining the position or not inter-viewing for it could be perceived negatively as a lack of ambition. Over the years, I found that I worked the long hours during the week and over the weekends in hopes that the appearance of my work performance and balance would appear as normal as possible. On more than one occasion, family and close family friends traveled to care for my daughter as I supported company initiatives and traveled (domestically and internationally). The thousands of dollars in costs for additional travel were our personal expense, not reimbursed by the company, and were often done without mention. Personally, I did not want my marital or motherhood decisions to be a topic in the decision-making rooms where I was not present. It was an attempt to control perceptions and my image.

Dwayne Hills Sr Photography

A smile and silly moment for Raelyn and Ava Holmes (2013).

Raelyn Holmes

Parents, Raymond and Anna Holmes, visiting me at Stennis Space Center for a Space Shuttle Main Engine Test (1998).

Mentorship/Sponsorship

How important was mentorship/sponsorship for your career? Have they been men or women? How was the relationship established?

Mentorship/sponsorship was important for me. However, it has not always taken the form or structure that I anticipated. I have participated in both informal and formal mentorship. In my early years before college, sponsorship was led and offered by women in my community. As my career began, both women and men have been my mentors and sponsors.

One relationship began as a result of a job interview. I was a Space Shuttle Flight Support Project Engineer at the time with plans to enter program management. I interviewed for a stretch position in a panel-style interview and, ultimately, did not get the job. However, the managers that participated in the interview were introduced to me through the process, and at least one of them was interested in me applying for another position. I got the next job, and it was the right position for me where my strengths were utilized, and I was exposed to many new areas of the aerospace industry.

Another mentor not only shared her personal career path lessons but her leadership style, home, and family decisions/adjustments.

There have been occasions where I perceived mentoring limitations due to race or gender, but there were still career- and life-impacting moments from those relationships as well. One relationship led to my exposure to emotional intelligence, a skillset that is like a gift that keeps giving.

Most recently, I have admired a formal mentorship program and mentor because he showed me that distance, age, gender, and race would not be a limitation with our mentor-mentee relationship. He followed program guidelines (as appropriate) but his authentic style was the true contributor to our success; that relationship was a two-way mentorship in which we are able to share professional challenges and learn from each other's experience. Lastly, I have appreciated the learning and championing that occurs amongst the diversity of co-workers and my friends.

Avoiding a Stall

How did you develop organizational savvy? And how did it help your career?

Organizational dynamics have been a focus in some of my higher education courses. However over the past 23 years in the industry, skillsets such as listening, participating in initiatives, paying attention to cycles, accepting my failures and wins, staying present for the "next" opportunity, and being conscious of "what" I can control or influence have contributed to how I developed or honed elements of my organizational savvy.

I have collected several principles over time that have been helpful in my workplaces. These are a few:

- Everybody has priorities. What do your actions say about your priorities? What does the organization's spending suggest about its priorities?

- This industry is small. You will meet prior team members and co-workers, again; either as a counterpart, supervisor, consultant, customer, or client. Consider that next meeting before you part ways.

- Emotional Intelligence, be mindful of your reactions. It is okay to have emotion, but not to be emotional.

- If not you, then who?

- Share something about yourself. Be authentic and build connections. It matters!

- When I am uncomfortable or upset, be aware of the emotion and focus on what my impact or purpose will be in the moment (for example, will I create a "safe place" for learning/expression, influence thought, or help them to be the best at who they are).

- When "Zooming," look into the camera to make eye connection with a person. Appreciate that there is possibly less posturing in this virtual environment at the moment. Focus on the positives.

As a Global Trade Manager, my organizational savvy helped me to set up a functional compliance team and processes that shipped products and exported technology compliantly and on time (Sales ~$160M, Earning Before Interest and Taxes ~$40M). It was a challenge; actually, it was an extremely challenging time with extensive work and growth, but unexpectantly, more frustrating because individuals within the organization were not working towards the same business goals, satisfying the contractual commitments and maintaining a positive customer relationship.

The part of the organization ultimately creating the "head wind" was working with misconceptions, unjustified fears, and larger reorganization objectives, elements that were counterproductive to the immediate needs of the business objectives. When I realized how committed this part of the organization was to their beliefs and tactics, I had to make some key choices because of the nature of the reporting structure. It was my supervisory chain that was creating the "headwind" versus being the "tailwind" or removing obstacles. It was the first time in my career that I had experienced this type of a misled destructive dynamics, it was emotionally disheartening and surely had an impact on my health.

The first decision was to accept it, be aware, and stay outwardly unemotional; "they" believed they were in the right and taking the appropriate actions for the company. The environment was full of triggers such as decision/path changes, resource reassignments, micro-aggressions, aggressive criticisms, and double standards that seemed to challenge my emotional intelligence and presence in the moment. It all seemed so strategic.

My next decision was to continue to build and maintain connections. My affirmations were clear. I would be authentically who I am and had been. I would be present and available for the business, process, and resource development. I would embrace all the company policies and requirements (for me and my business). I would continue to apply my experiences and best, and learn from the new experiences. I was in the muck and mire, but I knew that overall, the company cared about results, compliance, safety, and respect.

In the end our team performed and won a company-sponsored Leadership Award. The portions of the business that were "head winds" ultimately took other leadership opportunities in the company. There was a very distinctive positive leadership and culture change in our division following that period of time.

Powering On

Was there a significant event that changed your career trajectory and what was it? Have you ever cried at work out of frustration or anger? If yes, how did you handle that?

I started my professional journey as an intern and scholarship recipient for 3M Company. I was introduced to 3M Company, as a student at Tuskegee University when I began my engineering studies in the FASTREC (a summer program for incoming engineering students). The summer of 1992, my father had given me one goal in the FASTREC Program—"If you want to attend Tuskegee, you need to complete this program with a scholarship." Such simple words and a charge, but I had led a sheltered, provided-for life, and I did not know what was required to make it happen. I did know that I wanted to continue school at Tuskegee, so I figured it out. I made an appointment with the FASTREC Program Director and asked about scholarships and how to qualify.

The take-away was to get good grades, and scholarships would come. I took four courses that summer, finished with a 4.0 GPA, earned an opportunity to interview with 3M Company, received a merit-based corporate renewable scholarship (based on GPA performance), and received an internship offer. The 3M Company experience formed a great foundation for me. Not only did I earn "my spot," but the 3M Company culture was welcoming, encouraging, and innovative. I felt a responsibility to do my best and was only constrained by the laws of science and the precautions for safety.

A number of years later I accepted a job in aerospace with Pratt & Whitney, the technology opportunity was enough for me to move further from home, accept a lower salary, and venture into an unknown company culture. By this time, I was a Project Engineer with plans to become a Program Manager. My technical background served me well, but the ability to team, organize, critically analyze, multi-task, communicate, and document seemed to be at the forefront now. While there were many positive and rewarding memories, this moment was not one of them.

I had started estimating engineering work and preparing basis of estimates to support government contracts. I was excited and wanted my work to be the "BEST ESTIMATES EVER." I had been invited, and I was going to give estimates that were accurate and would be referenced for years to come (I laugh as I remember my excitement). A moment in the process came that all estimators on the team came together in a large conference room to review the basis of estimates and identify opportunities. Other than Flight Readiness Reviews or Anomaly Investigations, this was the largest non-technical collaboration event that I had been involved in. The environment was tense, which was hard for me to understand since there was no shuttle on the pad, tank loading for test, or a ruptured part sitting in a quality crib. As the large team read the estimate and justifications, the Financial Manager walked around the room and commented. She was one of the few female managers I had experienced in my time with the company, and I was proud of her accomplishments and knew there were untold stories/struggles.

Then, she said my name, which I did not know she knew, and I looked up ready for engaging discussion. There was no discussion though, she began to relentlessly criticize the estimates and write-ups that had been created and approved by myself and my fellow Engineering team. I was paralyzed. I had experienced constructive feedback in the past, but not a public work slaying with no form or structure; this was my time. The room was silent, I was silent and possibly reflective, but there were no specifics to reflect on. When she finished, I sat for a moment then quietly excused myself from the conference room.

I didn't know where to go. I had nowhere to go, so I went to the ladies' restroom. The ladies room had a lounge with cushioned chairs and vanity mirrors. I wondered, "Does the men's room include a lounge? Does the ladies room lounge exist for what I am about to use it for?" I needed to give myself a moment to collect myself, regroup after that moment of pride and loss, and allow myself to pivot into "whatever" was next. As I came out of the ladies' room, another manager was waiting for me. He apologized for her behavior, explained that she had been "different" recently and nothing in the work deserved that response. Tears fell in that hallway from my eyes. I had a renewed confidence in mankind and Pratt & Whitney at that very moment! And I went back into the conference room.

Lynne Hopper

Vice President, Engineering and Chief Engineer
Boeing Commercial Airplanes

About the Author

Lynne Hopper was named Vice President of Engineering for Boeing Commercial Airplanes in March 2019. She also serves as a member of the Boeing Enterprise Engineering leadership team.

In this role, she leads 12,500+ engineers across the world who design, produce, certify, and support Boeing Commercial Airplane programs and products. Hopper confirms products are safe, designs are efficient, and products and services meet customer needs. She also ensures that all products are producible and meet compliance requirements and that customers can safely and efficiently operate them.

As the Commercial Airplanes Engineering leader, Hopper is accountable for commercial product design practices and creating a learning organization focused on a strong Safety Management System.

Previously, Hopper was Vice President of Boeing Test and Evaluation, where she was responsible for laboratory and flight test operations in support of validation and certification of Boeing commercial and defense products. Named to that position in 2018, Hopper oversaw about 5,000 engineers, pilots, mechanics, and technicians who test and evaluate new Boeing aircraft and modifications and upgrades to existing aircraft, and who provide test support to Boeing businesses.

Prior to that position, Hopper served as Vice President of Engineering, Modifications, and Maintenance for Boeing Global Services, where she was responsible for fleet engineering services, modifications to aircraft, logistics products and support, Aircraft-on-Ground services, forward base deployment with operators around the world, and Maintenance, Repair, and Overhaul facilities in San Antonio, Cecil Field, and Shanghai.

Hopper also served as Vice President of Material Services for Boeing Commercial Airplanes, where she was responsible for commercial aviation's most comprehensive aircraft spare-parts sales, services, and distribution network, serving customers worldwide. Concurrently, she led the Commercial Aviation Services' data analytics team.

As Vice President of Commercial Airplanes Customer Support, Hopper managed the health of the world's largest fleet of commercial jetliners – more than 13,000 airplanes. She oversaw a complete set of Boeing technical resources, helping more than 800 operators maintain their fleets throughout the life cycle of their airplanes.

Hopper led the development of regulatory-delegated authorized representatives for commercial airplanes from 2004 to 2007 and has held leadership positions in 737, 747, and 757 airplane systems and in airplane safety and airworthiness.

She holds a bachelor's degree in Mechanical Engineering from the University of Utah and a master's degree in Mechanical Engineering from Stanford.

Questions and Answers

Personal Career Insight

What inspired you to choose a career in Aerospace and what has given you the most satisfaction during your career?

My father was my earliest inspiration. He was a civil engineer who loved to fly airplanes. He took me on business trips and let me fly part of the trip when I was only nine years old. He instilled in me that I could do anything he could do and worked on projects with me in the garage and on the sewing machine. He died when I was young, but in high school, I went on to compete in International Science Fairs and excel in math and science.

I applied for and received a scholarship at the University of Utah, which was a huge relief because I had limited funds for college. After I graduated, I proudly started at the former Douglas Aircraft Company as my first engineering position, and later transferred to Boeing.

I had the opportunity to do analysis and tests on many of our products. I have flown simulators and test flights on commercial, military, and rotorcraft products. I represented Boeing to our customers as the Vice President of Customer Support. And I had the opportunity to lead very large organizations and implement very significant changes like standing up our first FAA delegation system and consolidating all of our Customer Support operations in Seal Beach, Calif.

Work/Life Balance

Did you ever feel guilty for leaving work on time to attend a family event? Did you feel it reflected negatively on you at work? If so, how did you deal with that?

It's not just women who have families, and I think the current generation will demand more equality in their role as parents and work/life balance. With that in mind, I think all career women who are mothers deal with guilt frequently. There have been family events I have missed, and there have been business dinners I have left early or passed on completely.

Lynne Hopper

I pass tips to people who are managing both roles. One of the best things to do is to have a plan. I have long-term and short-term goals I track for work. I do the same with my family. I plan summer weeks to make sure we make the most of the time, planning s'mores on the fire pit, camping in the yard, vacations, baseball games, and family get-togethers. I even made multi-year look-ahead calendars and thought about the perfect timing for a cruise or a significant vacation. I didn't want to lose the opportunity when our kids were of any age. I also plan meals for the week.

On Sunday, I pull out the binder of our favorite recipes, choose a handful, and shop for necessary groceries. Then, on any given night, I could pick a recipe and have dinner on the table in short order.

Having backup childcare was critical especially when our kids were younger. With no family in the area, I had a couple of babysitters that I could call to bail me out when I got caught at work or left on a business trip.

These days, my husband has taken over almost all of the laundry and grocery shopping and dry cleaning and store returns. Having a strong understanding with your spouse or partner is vital for professionals everywhere, regardless of the industry.

Mentorship/Sponsorship

What activities have you engaged in that have helped other women achieve success in their aerospace careers?

I am a huge believer in the value of diversity and work hard to make sure the Commercial Airplanes Engineering team works for inclusion and equity. We have more work to do across the board, for instance, more people of color need to be represented.

One of the things that I'm very proud of is the five years I've served as the Executive Sponsor for the Society of Women Engineers (SWE) at Boeing. The first conference I attended in that capacity, we hired about a dozen female engineers. The last year, we hired more than 150 engineers.

It was a lot of hard work by an almost entirely volunteer team. We were able to get many of our male engineering executives to attend the conference, including members of our Executive Council—the senior-most leaders of the company. We revamped our presence at the conference, expanding the size of our booth and getting our subsidiaries to participate so talent could see the wide range of possibilities at Boeing.

We started a program with SWE to onboard women engineers who had left the workforce and reacquaint them with engineering, and then Boeing could utilize their great talent. And through all that I realized that even if I felt like the only woman at the table many times, there were others who were even less represented.

I also am willing to have an informational meeting with anyone who asks. That openness has given me the opportunity to meet some wonderful new talent who I can help in their careers. I learn from them and it helps me to be a better leader.

Avoiding a Stall

Were you ever presented an "opportunity" you declined and did it hurt your career? If not, how did you overcome any negative impact?

When I was younger, I was open to any career opportunity that came along. It felt good to be asked. At one point, I'd made the short list to leave engineering to go to contracts. One of the directors took me aside and asked if I knew how that may change my career path. He thought maybe I should consider staying in engineering. He suggested I talk to some leaders who had worked in engineering and the sales/ contracts side of the organization.

It was very enlightening. I stepped back and thought about what gave me joy at work, and what drained my energy. And I decided to get myself off that list, very graciously, and stay in engineering. Not only was this a better decision for me and my career, I also continued to get support from the three leaders I had met with to discuss careers. I now advise people who I mentor to use job changes and career decisions as a way to grow their network.

I recommend going out and talking to people who have jobs aligned with your aspirations and find out what it would take to be ready for a similar role. You can learn about the job and potentially create a sponsor at the same time.

Powering On

Have you ever cried at work out of frustration or anger? If yes, how did you handle that?

As a female engineer working in a predominately male business, there are many times I have been frustrated or angry. Interestingly, it was a female manager that made me angry enough to cry the first time. She wanted us all to join a phone tag-up every morning at 6 a.m. and daycare didn't open until 6:30 a.m.

The second time, I wept when the company handed out job offers to more than 150 female engineers at the SWE convention. I was so proud to stand in a room and see all these women who would be joining Boeing's engineering team. I knew they would make a difference, because the work we do makes a difference. What keeps me motivated is my passion to make a difference. I have a passion for Boeing and the people who work here. We protect, we connect, and we inspire.

Amelia Earhart once said that adventure was worthwhile in itself. We work in this profession because of the inherent benefit it provides for everyone.

Even as the world went through the devastation of the September 11th terrorist attacks, or now with both the 737 MAX fleet being grounded and COVID-19 impacting us and our customers, people are working so hard to do the right thing. Our focus on safety, quality, and continuous improvement will get us there. I know by taking care of each other, we will find further opportunities to keep the world flying safely and confidently.

Kathleen C. Howell

Hsu Lo Distinguished Professor of Aeronautics and Astronautics
College of Engineering, Purdue University

About the Author

Professor Kathleen Connor Howell is presently the Hsu Lo Distinguished Professor of Aeronautics and Astronautics in the College of Engineering at Purdue University. She earned her BS degree in Aerospace Engineering from Iowa State University and immediately joined Procter and Gamble Manufacturing Company in Kansas City, Kansas. A manager in consumer product manufacturing, Professor Howell was promoted to Department Manager in the production of industrial chemicals prior to her departure.

Professor Howell entered graduate school and earned her MS and PhD degrees in Aeronautical and Astronautical Sciences from Stanford University. As a graduate student, she also spent time at the Jet Propulsion Laboratory in Pasadena, California, a NASA research facility supporting robotic space and science spacecraft missions.

Professor Howell joined the faculty at Purdue University in 1982. As an engineering educator, she teaches courses in mechanics and spacecraft attitude dynamics as well as orbital mechanics including design and analysis for spacecraft missions. Professor Howell's technical research focus is astrodynamics in complex gravitational environments, and she leads an active research group developing methodologies for space mission planning and on-orbit operations.

She has successfully applied these methodologies to numerous NASA missions. Her contributions include mission planning and trajectory optimization, station keeping and maneuver design, low-thrust applications including small satellites, and the development of interactive visual capabilities for complex mission scenarios. As a principal investigator, she has obtained numerous grants and received various awards related to her research program as well as the recognition of her service as an engineering educator.

She served for many years as the Editor-in-Chief for the American Astronautical Society (AAS) Journal of the Astronautical Sciences; she is also a member of other editorial boards. Professor Howell is a member of the National Academy of Engineering, the American Academy of Arts and Sciences, the Celestial Mechanics Institute, and the International Academy of Astronautics. She is also a Fellow of both the American Institute of Aeronautics and Astronautics and the AAS. She is involved with various other organizations within the international aerospace and astrodynamics community.

Questions and Answers

Personal Career Insight

What inspired you to choose a career in Aerospace and what has given you the most satisfaction during your career?

My inspiration comes from a number of lines that converged. I always found math and astronomy and even the history of astronomy fascinating. But my real interest emerged with the thought that humans could go to these places... sometimes physically, but we could always send robotic spacecraft to these fascinating places. How is it possible? What did I need to learn to make that happen?

Of course, I was also just starting college when Apollo landed on the moon...it was real and I wanted to know how it was done. What did I need to study to learn it? Finally, it was also a time that women were not readily encouraged to go into engineering. My father was always forward thinking. As the General Counsel of his large company, he was a leader in hiring women lawyers at a time when they were rare. My high school counselors argued that I was better suited elsewhere. But my father told me that if I wanted to learn to send spacecraft throughout the solar system...do it. He encouraged me to find my passion and work to "be really good at what you do." In challenging financial times, he and my mother found a way to send me to a university with a strong Aerospace Engineering program.

One of the parts of my career that gives me the most satisfaction is in my role as a professor in the classroom and as a mentor to graduate students. I love to spark excitement and nurture students who also have a passion for the subject. It is also a real opportunity to encourage those that are hesitant to find their way in a field to which they aspire.

Work/Life Balance

Did the pursuit of a career impact your decision on whether or when to have a family? What was the impact of that decision?

My family is my highest priority. However, my spouse and children have been totally supportive over my entire career. My husband is an industrial engineer with a long-term career in his own field, and he has been my biggest cheerleader. He and I lived in two different cities most of the time while I was in graduate school. We made such decisions together and worked hard to be sure that we each could pursue professional opportunities. Our children were raised with both parents in careers and that was their "normal." We were both very involved parents, and part of the attraction of a relatively small community in Indiana was living very close to their school. They are great young men now and I think they are proud of their parents as we are of their accomplishments.

Mentorship/Sponsorship

How important was mentorship/sponsorship for your career? Have they been men or women? How was the relationship established?

Mentorship and advocacy are keys to success for every career. I have participated in activities to both mentor and advocate for women faculty in STEM. But one thing that I learned from my male colleagues is that many of them did not even recognize the mentoring they received along the way…sometimes it is subtle, but so important, almost as if it is taken for granted by both informal mentors as well as mentees. Many of the young women faculty do not receive the same access to such networks and the frameworks for development.

As I look back, there were many men and women who were key to my success. Some involved only one significant interaction, some were longer term. Almost all originally occurred by accident initially. Notably, during my academic career, there were two Deans (one male, one female) who were really both mentors and advocates at different times. The advice I might give is to be open to learning from more experienced folks and to acknowledge their assistance.

Avoiding a Stall

Have you ever taken a role you were not excited about but had to show you were a "team player"? What was the outcome?

As a university professor, my role is quite broad. I was also the only female professor in my department for many years and actually one of only a handful of female faculty in all of the College of Engineering over a significant arc of my career. As such, I saw (up close) the challenges and small steps as the number of women in engineering slowly climbed over time. Those facts alone gave me visibility and I was called upon to participate in a variety of activities.

As I look back, these early activities offered opportunities that I did not anticipate. I met and interacted with more faculty across campus and more campus leaders than other faculty at my rank. Those interactions over time have given me some of my strongest advocates. Yet some of these roles offered challenges to me personally as they were not in my areas of strength. For example, later in my career, I have been offered administrative positions that are critically important in a university setting. I have served in such capacities for short intervals as a recognition that these are opportunities to serve and a chance to effect change. But I always prefer to come back to my roots engaging with the students.

Powering On

As you hit career obstacles, what motivated you to keep going?

I think we all hit obstacles. As I was an assistant professor on the tenure track, I kept going by realizing that it is more important to remain true to myself, not to remake myself to respond to external pressures. I decided to work on problems that intrigued me and stick to the path that brought me to that point. Some self-awareness and self-confidence that I did have other choices that could lead to equally exciting opportunities was hugely reaffirming.

Some other types of obstacles involved issues that I could not correct myself. Those are challenging because "just working harder" is not the answer. Recognizing when I needed outside assistance meant an honest assessment of my options. Getting over or around an obstacle might be enabled from new perspectives or more creative approaches. But it might also imply a willingness to say "enough" and move in a new direction. My original career path right out of college was management in industrial manufacturing. A self-assessment shifted my focus to a career in academia. In fact, it is a much better match for me!

Miyuki Humer

Program Manager
General Atomics

About the Author

Born and raised in Japan, Miyuki Humer studied Machine Intelligence and System Engineering at Tohoku University, before traveling to the United States and earning her Master's in Mechanical and Aerospace Engineering from the University of California, San Diego.

Miyuki began her career as an aero/thermal engineering intern at Hamilton Sundstrand Power Systems. During her more than eight years at the company, she served in a variety of roles, including senior mechanical engineer, system engineer, project engineer, and system integration engineer.

Transitioning into project management, Miyuki joined Goodrich Aerostructures Company, where she focused on overseeing project schedules and costs. She next moved to General Atomics Electromagnetic Systems Group, managing engineering changes and configuration. Today she is a hardware project manager, responsible for the on-time delivery of hardware for government production contracts.

Miyuki has three daughters and enjoys being a full-time working mom. She also enjoys many outdoor activities, such as traveling to national parks, camping, hiking, surfing, and skiing with her family.

Questions and Answers

Personal Career Insight

What inspired you to choose a career in Aerospace and what has given you the most satisfaction during your career?

I was always fascinated with flying since I was little, and I wanted to be a pilot. Due to my bad eyesight, I had to give up that dream, but still wanted to work on something related to flying. My most satisfaction was when I flew on a Boeing 787 for the first time. I had worked on developing and certifying the Auxiliary Power Unit (APU) and other systems for Boeing 787, so I felt like the airplane was mine.

Work/Life Balance

Did you ever feel guilty for leaving work on time to attend a family event? Did you feel it reflected negatively on you at work? If so, how did you deal with that?

Yes, I feel guilty when I have to leave work earlier to fulfill a family obligation, and I am worried about if it is viewed negatively by my boss and/or co-workers. All I can do is to catch up on work after hours when I can and keep up my performance while remembering that the work performance should be evaluated by the quality of the outcome, not quantity of the time spent. I also make sure to show my empathy to my co-workers/subordinates so that they do not feel guilty when they have to leave early due to a family matter

Miyuki Humer

Mentorship/Sponsorship

How important was mentorship/sponsorship for your career? Have they been men or women? How was the relationship established?

I think it is important to have mentorship for your own development and your career advancement. Unfortunately, I never had a chance to be in the formal mentorship program, but I have met several individuals I respect. They include men and women who were my boss, indirect boss, and co-workers. I learned a lot from them technically and nontechnically. I would observe how they presented themselves in meetings and how they resolved conflicts and overcame challenges. I also asked them to give me feedback on how I was doing in those situations. I have had many benefits from mentors for developing my professional knowledge and attitudes.

Avoiding a Stall

Have you ever taken a role you were not excited about but had to show you were a "team player"? What was the outcome?

I was not too excited when I was told to temporarily support the system-engineering group while I still belonged to the mechanical/testing group. It was completely outside of my comfort zone as I had no knowledge of software design and certification, logics, and fault detection systems used in APUs. However, this is one of the best things that ever happened to my career expansion and to me. It was great fun to learn a new engineering world that I had not much education or work experience in and to meet a different set of working group people to learn from. Rhonda was one of my great working partner/mentor at that time who taught me patiently how system and integration engineering worked. I expanded my interest and work experience outside of mechanical/test engineering and that helped me to advance my career to the next step.

Powering On

Was there a significant event that changed your career trajectory and what was it?

I would say definitely having children is a significant event that changed my career path. As my children get older, their demands have changed where I need to be closer to their physical location (i.e., school) to drop off and pick them up at certain times, drive them to different activities, and so on. I had to find a job within a 10-minute commute from home and their schools. Luckily, I found the current position that fits with that requirement, but I had to leave the commercial aerospace industry. I do not see this as a negative impact on my career at all, rather considering it my exploration period of what and how other industries work.

I enjoy learning different things every day from work and spending more time with the most demanding phases of my children at this moment. I am sure this will come to end and will move on to the next phase where I can look for other opportunities that will fit into my lifestyle then. To me, it is important to have both—a professional work life to achieve satisfaction from the outside world and a personal life to have children to love and get personal satisfaction. It is hard to balance well between these sometimes, but that is the part of fun and challenges I enjoy. I hope more and more women can have many options and society will improve more to support such women in the work environment.

Jamie Korman

Lead Engineer
Collins Aerospace

About the Author

Ms. Jamie Korman is the Smart Factory lead for the Digital Operations department in the Operations Modernization division at Collins Aerospace. She leads a technical team that supports Industry 4.0 and Connected Factory initiatives. She is currently in the Development Engineering Leadership for Talent Advancement Velocity (DELTA-V) program for aspiring fellows where her discipline is Industrial Engineering with a focus on Digital Operations and Data Analytics for Manufacturing. She serves as the Dean of the College of Industrial Engineering for the Collins Aerospace Technical University (CATU), where she is implementing Industrial Engineering curricula to provide internal educational opportunities.

Ms. Korman previously served as a Research and Development Engineer at the Applied Research Laboratory at The Pennsylvania State University, which is a University Affiliated Research Center (UARC) for the Navy. As a trusted agent, she served in various positions with growing responsibility. In this role, she developed algorithms for Condition-Based Maintenance efforts for the US Army. In addition, she was a Principal Investigator responsible for the development of forecasting evaluations for supply chain logistics and developing simulations to evaluate optimization models for Naval Supply Systems Command.

Prior to that, she served as an Industrial Engineer for the American Clip Company (ACCO) where her focus was on continuous improvement. She led events focused on setup reduction. She earned her lean certification through MANTEC during this time.

Ms. Korman earned her Bachelor of Science degree in Industrial Engineering with a focus on Manufacturing Systems from the Pennsylvania State University.

In addition, she earned her Master's degree in Data Analytics and a Graduate Certificate in Applied Statistics from Penn State.

Questions and Answers

Personal Career Insight

How did you decide between a leadership vs. technical career track?

The short answer is that I have not decided which career track to follow. I realize that I am in a fortunate position to have the opportunity to choose, but the reality is that I truly enjoy the challenges associated with both options. For me, this decision does not have to be a linear one, rather I think it's important to be a forever learner so that regardless of position, leadership or technical, I can execute for the greatest needs in Aerospace.

Currently, I have the honor of leading a technically diverse team of extremely talented contributors. In that role, I provide leadership, guidance, and resource coordination, as well as support in job performance and goal attainment. I particularly enjoy the mentorship aspect that's associated with being a technical team leader. I like being able to share my knowledge and expertise in the fields of Data Science and Industrial Engineering with the team, but not nearly as much as I enjoy learning from them. I am amazed at the unique level of expertise each member brings to the table. As true with all aspects of life, each member of the team will continue to grow and mature by learning through each other's individual experiences and technical accomplishments.

Prior to my current position, I functioned as an individual contributor, where I initially expressed my desire to remain technical. I thought that I needed to stay in that role so I could have time to grow my skills prior to navigating down the leadership path. After several years in that position, I was nominated by leadership for the Development Engineering Leadership for Talent Advancement Velocity (DELTA-V) program for aspiring technical fellows. This opportunity can lead me back to a technical career track in the future.

My advice, if you are new in your career, focus on the areas that get you excited. Find your technical passion and focus on it. Leadership opportunities will come, and when they do, accept the challenge and think of it as an opportunity for professional growth. While in a leadership position you will, as I have, develop a greater appreciation for all the technical work and synergy that is required to drive large technical initiatives. Eventually, I think the path of leadership or technical will be clear for me, but for now I am enjoying pursuing both.

Work/Life Balance

Did you ever have to make a move to advance your career (within your company or changing companies) that impacted your family life, and how did you balance the two?

In 2017, I accepted a position at Collins Aerospace that required my family to relocate. Prior to accepting the position, I was introduced to a woman at Collins that had recently moved her family for her career. She graciously shared her lessons learned and overall experience with family/job relocations. Speaking with her and hearing her story was inspiring and helped prepared me for the move ahead. Also knowing I had a friendly face at the new job helped put me at ease during the transition.

Jamie Korman

The three months between applying to starting were a complete whirlwind. We sold and bought a new home and relocated our family from Pennsylvania to South Carolina. Moving with a young family, I had to be thoughtful about my husband's and children's needs. My husband's hard work in his career paved the way for his company to allow him to work remotely. The fact that he didn't have to start a new job at the same time was extremely helpful. We moved three days prior to my oldest daughter starting Kindergarten. My husband took on most of the administrative tasks of supporting her transition to school and since he was working from home near the school, it put us at ease that he was close by in case of emergency.

We also had to balance moving away from an established friend group and family support to where we didn't know anyone. We were fortunate to move into a neighborhood where we immediately made friends that became a huge support system for us. I think having a network of friends and family is key to any dual working family's success, especially when relocating. This new network helped us find new doctors, daycares, to even grocery stores.

I also had to be realistic with the expectations I put on myself. I recognized I needed to evaluate all current demands on my time and outsource or limit where possible. I reduced my educational workload from my master's program by only enrolling in one class that semester. In addition, I hired cleaning help and grocery services to shop for me. Meal preparation and planning paved way to my husband cooking more dinners and helped keep my favorite tradition of family dinner intact. Reducing these previous demands allowed me time to focus on my family, new location, new position, and keep pursuing my masters. This was also the first time in my life I learned I don't have to try and do it all and that it's ok to have help. Having a supportive spouse, family, friends, neighbors, and a network of strong female mentors was paramount in the decision to balance my career move with my family.

Mentorship/Sponsorship

How important was mentorship/sponsorship for your career? Have they been men or women? How was the relationship established?

Mentors have had a huge impact on my career. I've been fortunate enough to have both male and female mentors and advocates. Some of the relationships were established formally, some were ad hoc, and others were established through networking. I am extremely fortunate that my current mentor asked me if she could mentor me when I started at my current company.

Coincidentally, a male colleague at my previous employer had connected me to this woman via LinkedIn prior to my start date at the new company. I pursued talking with her, but I never thought to ask if she would mentor me. After getting to know one another, she asked if she could be my mentor. I felt truly honored. This is a woman that was extremely knowledgeable, accomplished, and well respected within the organization. It has been helpful and reassuring to hear about her successes, challenges, lessons learned, and knowledge gained throughout her career. It especially great to have a trusted confidant to talk through work and personal successes and challenges.

In addition to a formal mentor, I also had an informal male mentor. I met him at a project kick off meeting. He was a Fellow and recognized my skill set could help solve a challenge a site was facing. He mentored me through the project and helped to connect me to the right people to showcase the work I had done. He even worked with my manager to ensure that I was nominated for a program that helps grow technical talent on the path towards fellowship. This program provided me with a formal male steward that is currently guiding me on the path to become a technical fellow. This mentor provides guidance to ensure I have the breadth of knowledge to earn a nomination to become a fellow. This mentor connects me with people across the organization I might otherwise not had the opportunity to meet. I likely would not have met my formal steward if it wasn't for the informal mentorship prior to it.

These mentors have also connected me to other mentees that I have been able to mentor formally and informally in my career. I have found that even one conversation can positively change a person's perspective. Knowing how impactful these people have been to me, I want to have a similar impact to the mentees I have the privilege to work with.

Avoiding a Stall

How important is an advanced degree?

I've seen people be successful without a formal degree, and I have seen PhDs flounder with anything outside their PhD thesis expertise. So I do not think a degree is what defines success. However, with that insight, I still think it's important to always be learning and expanding your knowledge base, whether that be via a formal advanced degree or some other educational avenue. I believe if you are going to pursue an advanced degree, make sure it is something that you will enjoy and that will provide a return on investment.

For me, starting an advanced degree in Applied Statistics was after a colleague had introduced me to the field of Data Science. At that time, it was not a well-known field of study. I enjoyed working with data and an advanced degree gave me the understanding of how I could leverage data in new and exciting ways. Researching job profiles of Data Scientist also assured me that my investment into an advanced degree would have a monetary return in addition to career satisfaction.

During my master's studies, I found an internal job posting for an R&D Engineer that was more focused on Data Analytics. To be honest, the job felt above my skill level. However, the hiring manager appreciated my focus of study and was willing to hire me on the skills I was seeking to grow with an advanced degree, not necessarily the skills I currently had. That hiring manager, trusting in my desire to learn and grow, has shaped how I interview and hire today. Although I don't necessarily require that someone has an advanced degree, I like to see people pursuing continuing education in one form or another.

While working towards earning a Master's in Data Analytics, I again found a Data Scientist posting for a position at Collins Aerospace. Having that advanced degree was pretty much a requirement. Honestly, I don't know if they would have considered me for the position without the degree. I think pursuing an advanced degree provides a focused area of study. Being a forever learner makes you a better resource overall, and advanced degrees are a great way to grow technically.

Powering On

Was there a significant event that changed your career trajectory and what was it?

Having a child was very significant to my career trajectory. I had my first child a little over a year after graduating with my bachelor's degree. I was working my first job out of college and realized that if I was going to work outside the home, I had to be excited about it. Although I appreciate what I learned at my first job at the American Clip Company (ACCO), after previously having internships in Aerospace, it simply was not as exciting to me. As a result, I focused on my advanced degree, and I found a job as an R&D Engineer at Penn State, focusing on applied research for the Department of Defense (DoD).

The R&D Engineer position allowed me time to grow technically at a rate that might not have been possible had I followed a traditional corporate path at a large company. Between working full time and earning my masters, I was focusing on the field of Data Science 10 to 12 hours a day. I became steadfast in focusing on my career and striving for the right next role that would provide a strong career path and a good location for my family. Ultimately, I think I am far more successful today because of it.

Having children, especially daughters, inspired me to be stronger, harder working, and driven to inspire them that it's possible to succeed in both family and professional life. I also want to make every moment away from them count, because when I am with them, I don't want to be distracted by work. I am grateful my daughters and son will see two working parents come together and share all aspects of raising a family that break traditional gender stereotypes. Having my children inspires me to be the best I can be both at home and work, which ultimately has had a positive impact on my career trajectory.

Lorraine Martin

President and CEO
National Safety Council

About the Author

Lorraine Martin is the President and CEO of the National Safety Council, a nonprofit with a century-long legacy of eliminating preventable deaths at work, at home, and on the road. Lorraine is passionate about her commitment to helping people live their fullest lives, free of preventable injuries.

Lorraine has over 30 years of experience leading and developing global and international businesses, including as a Fortune 500 senior executive, with a successful track record in both civic and corporate roles.

As part of the Lockheed Martin leadership team, where she served as Executive Vice President and Deputy of Rotary and Mission Systems, Lorraine provided vision and leadership to a $16B global security and aerospace division charged with research, design, development, manufacturing, integration, and sustainment of advanced technology systems, products, and services. She led a team of 34,000 global employees in the United States of America, Canada, Mexico, Taiwan, Japan, the United Kingdom, Germany, Poland, the Netherlands, New Zealand, and Australia with operations in more than 75 worldwide facilities.

Over her career she has led global aircraft and complex system development and manufacturing, always with a focus on safety for the employees and for those who used the products, often in high consequence environments. Among her top achievements, Martin led the largest defense aircraft program in history: the F-35 Lightning II, a stealth fighter. She also spearheaded the successful operational and cultural integration of Sikorsky, a global leader in helicopters.

Lorraine is also Co-founder and President of the Pegasus Springs Foundation, a nonprofit organization dedicated to providing opportunities for educators, students, and community members to collaborate on learning models. She is an enthusiastic advocate for school, community, and national resource engagement.

As a proud champion for advancing women and girls in science, technology, engineering, and mathematics (STEM), Lorraine was recently named among STEMconnector's 100 Corporate Women Leaders and frequently lectures on core issues related to the cause. She has worked with numerous organizations in support of this mission, including Girls Inc., Girls Who Code, and Blue Ribbon Schools of Excellence, where she also served on the Board of Directors. In addition, she has served on the Board of Directors for INROADS and Big Brothers, Big Sisters of Orlando.

Lorraine serves on the Board of Directors and Audit Committee for Kennametal, a $2.4B global materials science firm that serves customers across aerospace, earthworks, energy, general engineering, and transportation.

Lorraine is honored to have served as an officer in the U.S. Air Force, where she held various leadership positions for software-intensive technology and development programs. She earned an M.S. in Computer Science from Boston University and a B.A. in Computational Mathematics from DePauw University.

Questions and Answers

Personal Career Insight

What was the coolest thing you experienced in your career?

I have had so many opportunities over the years to witness the amazing impact dedicated teams can make. Perhaps one of the most rewarding experiences, though, came during my time leading the F-35 program at Lockheed Martin. The F-35 is a stealth fighter aircraft designed and built for three U.S. services and numerous allies around the world. One operation it must perform is landing safely in the middle of the ocean on an aircraft carrier.

Martin with the Air Boss on the USS Nimitz in November 2014 for the F-35's first carrier landing.

The team had designed the F-35's unique arresting hook, which is the mechanism hidden in the belly of the aircraft that deploys when the aircraft is about to land and come to a stop on the carrier—a move that is called a "trap." There are four wires the arresting hook can hit and latch onto upon landing; the third wire is the optimal cable to hit.

After several months of painstaking work at the test range on land, it was clear that the arresting hook was not reliably catching any wire. The motion dynamics of the arresting hook were off.

With these failed tests, several senior leaders thought we may need to redesign the entire mechanism and begin from scratch, but the team's engineers felt differently. They believed quite strongly that, if given a bit of time, they could modify and fix it. We just needed a more sophisticated understanding of the dynamics, coupled with extensive modeling, to get it right. I trusted them, and they got back to work.

With some slight modifications and extensive testing at the range, we were ready to join the USS Nimitz, off the coast of San Diego, and demonstrate carrier operations with the F-35. We arrived at the Nimitz via helicopter, landing in the middle of the ocean with nothing in sight except blue water—until two F-35s appeared on the horizon. I joined our team of engineers on the catwalk above the flight deck to witness that historic first trap.

I will never forget that moment or that feeling of pride and joy: the arresting hook deployed from the first aircraft and hit the third wire, completing a picture-perfect trap on its first try. After so many traps, each wire is replaced, and the U.S. Navy gave me a piece of that "three wire"—a cable about 3 inches in diameter. I also have a framed copy of the USS Nimitz flight plans for that day. These are still some of my most cherished professional mementos.

In every career, there will be opportunities to witness moments that mark the culmination of a tremendous amount of innovation and work. In those moments, it is critical to remember what made them possible. For me, that day was possible because of an amazing team demonstrating perseverance, trust, and a commitment to excellence. To work with such a team was gratifying on a level I still struggle to describe. By the time we stood on that catwalk, I had no doubt that the plane would hit the third wire. And being there with the members of the team that made it possible was an honor I will forever treasure.

If you could give early-career aerospace women one piece of advice, what would it be?

Do not let someone else's baggage weigh you down. Putting others' issues on your shoulders just slows your progress—their baggage is not your weight to carry. Let me explain.

As an early-career aerospace woman, while I was still in the U.S. Air Force, I was the recipient of someone else's unconscious bias. I was certainly not alone in having such an experience. This was in the early 1980s, and while there was a growing number of women in STEM fields, women most often found themselves the only woman in the room.

One afternoon, I found some quiet time to sit in a conference room to create a technical presentation for an upcoming meeting on early developments in artificial intelligence. The most senior leader stopped in to see what I was doing. After I explained, he shared with all sincerity that he felt the skills I was developing would serve me well when I left the U.S. Air Force and was interacting in Parent Teacher Association (PTA) meetings. He offered that the other mothers would not have the skills I had built, and such skills would serve me well at PTA meetings.

I do not want to take anything away from the important work of PTAs, but I was appalled and disappointed that my most senior leader did not see or recognize my aspiring technical presentation skills that I worked so hard to build. I knew his intentions were positive—he was speaking with pride, but he did not encourage or envision any other applicable use for my skills. It was at that point that I made a choice not to own something that was not mine to own. I responded with a simple, "Thank you."

His words truly were a reflection of his experience; they were not a reflection of who I was, my aspirations, or the potential my career held. It was powerful to realize that I could hear his perspective... but that did not mean I had to own it. I did not have to carry it with me.

Today, I would have had more to say, and I encourage everyone to speak up when inappropriate assumptions are made, or inequities are present. Whether done on the spot or at a moment best for you to have the impact needed, expressing your concern can truly shift the situation and be a gift to a colleague—if they are open to receive it.

Bring your best to all you do, hone your craft, and lead with integrity. When faced with someone else's blindness or ignorance, help to set it straight, but know that their bias (conscious or unconscious) does not define you. Only you do that.

Work/Life Balance

Did you ever have to make a move to advance your career (within your company or changing companies) that impacted your family life, and how did you balance the two?

I was fortunate because I never had to leave my company to expand my contributions and continue to grow. Lockheed Martin has a diversity of products, business units, and operations around the world, so I did not need to look elsewhere to experience a dynamic and rewarding career. That said, I did physically move about every three years.

It is a gift to be able to say "yes" easily to new opportunities and challenges when they are presented to you. Being able to say "yes" will open the door to a multitude of experiences and chances to grow and learn. Thankfully, I never had to say "no" when opportunities presented themselves, so I accepted roles that took me on a journey I could never have imagined at the start.

My husband had a career that was best served by not moving. As such, he stayed in one location while I created a second home in each of my new job locations. Like many executives, air travel became part of the commute. Clearly, this is not a solution for everyone, but it has worked for us for 30-plus years. I realize that not everyone has the ability to say "yes" without having to uproot family members as well—but saying "yes" certainly helps to expand opportunities. Of course, everything changed in 2020 with the onset of the pandemic. Despite the well-documented challenges, weathering the virus from home has brought one major silver lining: the ability to spend more time with our loved ones. When my mother needed surgery this fall, for example, I was able to be by her side for her recovery, while taking the proper precautions to help keep both of us safe.

Lorraine Martin

Martin with her mother.

I often share that when you do move for your job, especially if your home base stays elsewhere, you must put down some real roots in the new location. It is important to make sure your team feels like you are part of their community. That is a critical connection, even if it is clear that the company is likely to shuffle the deck and ask you to move yet again sometime in the future. Looking back now, I think the adventure has enriched our lives, and my husband and I have also made lasting friendships with people around the country.

Mentorship/Sponsorship

If you had a sponsor to reach executive levels, how did you meet that person? Was the person assigned to you or was it a boss that became your sponsor? How did this person help your trajectory?

To me, sponsors and mentors are different and not necessarily interchangeable. A sponsor is the person at the table, advocating and going to bat for you, and asking, "Have we considered Lorraine for that leadership role?" A mentor is the coach on the sideline who ensures you are getting the best chance to make the most of your opportunities. A sounding board and a mirror, depending on the need. Not everyone may see the roles that way, but to me, they are both impactful and uniquely different.

In my career, my sponsors tended to be senior executives throughout the organization who saw my potential and advocated for me to their peers. It was never an assigned role; it was an organic relationship born out of their experiences with me and my contributions to the organization. Often a sponsor became aware of me through my involvement in special assignments that brought exposure to senior leaders not in my organizational chain. Offering to be on special teams that take on enterprise-wise challenges can be game changing.

At one point in my career, a sponsor advocated for me to move into the aircraft division. I had little to no experience with aircraft and their manufacturing. He knew of an opening and approached the head of that division, encouraging him to consider me. I have since learned that there was quite a bit of cajoling required. Without this initial support, I would not have gone on to lead the largest defense aircraft program in history. That sponsor's support changed my entire career trajectory forever. I would be remiss not to take this opportunity to say again—thank you!

The experience underscored the importance of the sponsor's role. It is so impactful to have people throughout your career who see your various talents and convince others to take a chance on you, especially for assignments beyond your previous experiences. Then you must do all you can to not let them down.

Avoiding a Stall

Have you ever taken a role you were not excited about but had to show you were a "team player"? What was the outcome?

Very early in my career, I was working in a research department responsible for a wide array of software projects. The business area leader asked me to leave that work and take on a novel radar program that used commercial radio waves to track moving objects, such as aircraft. There was not much of a market for such technology, and it had many doubters. Most concerning to me, however, was my lack of experience in the area. I had never worked with hardware—I had been a software geek. I did not know anything about radar technology, and it was a smaller project compared to the portfolio of software projects I was leading.

The radar was one of the business area leader's passion projects, though. He believed this technology could change the world. To me, it felt like I was moving both backwards and into foreign territory. I felt out-of-place and certainly out of my comfort zone. However, I quickly realized that he was giving me a unique opportunity to learn from him while simultaneously positioning myself for future opportunities that I could never have imagined leading.

The experience taught me a valuable lesson about trusting yourself and being curious. I did not have a background in the technology I was overseeing but I needed to trust myself as much as he trusted me with a project he valued greatly. He showed me that I did not have to limit myself to only working on areas where I had deep domain knowledge. I could lead a wide variety of technical programs, especially if I took the time to ask questions, learn from those around me, and leverage experts.

We won an innovation award for that project and it set me on a track for new and exciting opportunities. If I had passed on what looked like an unlikely next assignment, I may have had a very different career. Taking this alternate path gave me the confidence to try things that are completely outside the realm of my training, experience, and comfort.

Powering On

Have you ever cried at work out of frustration or anger? If yes, how did you handle that?

Every professional has had a handful of emotional moments at work, and I am no different. One distinct situation comes to mind.

A customer experience at a later point in my career caused some strong on-the-job emotions. Customer relationships are extremely important to me. I pride myself on knowing my customers' needs, listening to them at all times, and partnering to bring value. Program management is a "contact sport." You have to lean in if you are going to get it right, even if you might get bruised from time to time.

One day, a critical customer did get into my head. During a routine status update call, the team's engineers briefed our customer on their work with suppliers to resolve a persistent issue, and the customer started to accuse my team of blatantly misleading him. My team was being accused of things that were factually inaccurate and inappropriate.

I calmly told our customer that he was crossing the line and warned that I would end the call if he persisted—which he did. So I ended the call, as promised.

I was shaken, literally shaking with emotion. I asked the team: "Did I just do that?" They shared that I had warned him three times. Before taking a moment to myself, I made it clear that I fully supported them and had no doubt they were doing everything to resolve the issue. Their integrity would not be questioned.

I was so furious, there were tears as I shut my office door. The experience will stay with me always. While I knew that this customer had the best interest of the program at heart, I did not agree with his approach and could not allow it to impact my team.

Assuming goodwill is ultimately extremely empowering because it allows you to find solutions while changing the dynamic from something challenging to something positive. There are very few individuals in the world who truly wish you direct harm. That said, no doubt you will find individuals that behave inappropriately. Hone your techniques to address such behavior. Then regroup, assume goodwill, and lean in to find solutions.

ANOTHER TIME At the midpoint of my career, my team was in the midst of a bidding process for a large program to put sophisticated software into aircraft. Another player with an inherent cost advantage was bidding against us. Even though we knew we had all the right resources and personnel to win the bid, we were concerned.

I shared our concerns about our competitor's cost advantage with the customer, who assured me that it would not be a determining factor in awarding the job.

The call came in from a senior official late in the day. We had waited anxiously for hours to hear whether we had won. We had lost. To make matters worse, we learned that the cost differential *had* been the determining factor.

I shut my door and let the emotions wash over me. It was all there—sadness, frustration, anger, and perhaps most prominently, the sinking feeling that I had let down the team. With a deep breath, I went to talk to the team. There was much yet to understand about the loss, but for the time we needed to support each other and look to the future. I shared that I could not have been more proud and honored to work with each of them. The tears filled my eyes.

That moment was among the most challenging of my career. It also taught me the power of learning from loss, picking yourself and those you lead up, leading a team to the future, and finding the silver linings in any situation—they are always there, so keep looking.

Lourdes Maurice

DLM Global Strategies, LLC

About the Author

D r. Lourdes Maurice has served on the Advisory Board of Boom, a start-up company working on a supersonic passenger commercial airliner, since 2017. She is also the owner of a consultancy, DLM Global Strategies, LLC., focused on aerospace, environmental issues, and international relations.

Dr. Maurice was selected as Executive Director of the Office of Environment and Energy in March 2011. She served in that post until March 2017. In this capacity, she was responsible for developing, recommending, and coordinating national and international standards, policy and guidance, research and studies, and analytical capabilities on aviation environmental and energy matters. In addition, she represented the United States in the International Civil Aviation Organization (ICAO) Committee on Aviation Environmental Protection. Dr. Maurice oversaw efforts to establish regulations and standards, provide guidance and technical assistance for Federal Aviation Administration (FAA) compliance with applicable Federal environmental and energy statutes and regulations, and develop analytical tools and metrics to assess aviation environmental impacts. She also oversaw policy, applied science, and technical research programs to address aviation's environmental and energy issues.

Dr. Maurice joined the FAA as Chief Scientific and Technical Advisor for Environment in October 2002. In that capacity, she served as the agency's technical expert for basic and exploratory research, and advanced technology development focused on aircraft environmental impacts and its application to noise and emissions certification and policy, and the application of alternative fuels to mitigate environmental impacts. Dr. Maurice also founded, managed, and provided agency technical leadership for the Partnership for Air Transportation Noise and Emissions Reduction Center of Excellence.

Prior to joining the FAA, she served as the Air Force Deputy, Basic Research Sciences and Propulsion Science and Technology in the office of the Deputy Associate Secretary of the Air Force for Science and Technology. Dr. Maurice also worked at the Air Force Research Laboratory's Propulsion and Power Directorate from 1983 to 1999, planning and executing basic, exploratory, and advanced development propulsion science and technology programs, focusing on state-of-the-art aviation fuels and propulsion systems. Her areas of expertise include pollutant formation chemistry, combustion kinetics, hypersonic propulsion, and aviation fuels.

Dr. Maurice received her BS in Chemical Engineering and MS in Aerospace Engineering from the University of Dayton in Dayton, Ohio, and her PhD in Mechanical Engineering from the University of London's Imperial College at London, United Kingdom. She is also a distinguished graduate of the National Defense University's Industrial College of the Armed Forces, where she earned a MS in National Resource Strategy.

Lourdes Maurice

Dr. Maurice has served as Lead Author for the United Nation's Intergovernmental Panel on Climate Change (IPCC) and the National Academy of Science and National Research Council. She was recognized as a contributor to the Nobel Peace Prize awarded to the IPCC in 2007. She is an Associate Editor for the American Institute of Aeronautics and Astronautics (AIAA) Journal of Propulsion and Power and was on the Editorial Board of the International Journal of Aeroacoustics. She has authored over 100 publications and is a 2003 Fellow of AIAA.

Questions and Answers

Personal Career Insight

What inspired you to choose a career in Aerospace and what has given you the most satisfaction during your career?

I did not so much choose a career in aerospace as it chose me. I was a Chemical Engineering major at the University of Dayton in Dayton, Ohio, and my intent was to work in the cosmetics industry. I had my eye on Procter and Gamble, down the road in Cincinnati. But I had an interview with Herb Lander at the then Air Force Propulsion Laboratory in Dayton (it is now part of the Air Force Research Lab, AFRL). It was 1982 and we were going through an energy crisis. Herb was very passionate about alternative jet fuels (jet fuels made from shale oil, tar sands, and coal liquids at the time). I felt compelled to work with him. The energy crisis soon became an energy glut, but I stayed in aerospace.

I earned a Master's in Aerospace Engineering. I worked on ramjets, hypersonic propulsion, and other advanced propulsion concepts. The Air Force sponsored me to go to the University of London's Imperial College to work on a PhD. My research was focused on the chemical kinetics of jet fuel combustion (my PhD is in Mechanical Engineering). In London, I developed a social conscience through interactions with my tutor, Peter Lindstedt, and students from all over the world. We talked about how researchers had a duty to try to influence policy in a data-driven manner. When I came back to the U.S., I quickly found my way to Washington and started working on science and technology policy at the Pentagon.

Eventually, I landed at the Federal Aviation Administration working on environmental issues. I chose to do that after the attacks of 9/11 because I felt addressing societal issues, particularly climate change, was essential to continued peace and prosperity. At the FAA, I came back full circle to my first engineering job and launched an effort to develop Sustainable Aviation Fuels (SAF). Being part of the core group of folks that gave birth to this industry, which is so essential to aviation becoming carbon neutral, has by far given me the most satisfaction during my career. Herb Lander passed away a few years ago, I hope he was proud of what he started. He was an amazing man and a leader in giving women opportunities in aerospace.

Work/Life Balance

Did you ever have to make a move to advance your career (within your company or changing companies) that impacted your family life, and how did you balance the two?

Did the pursuit of a career impact your decision whether or when to have a family? What was the impact of that decision?

I did not realize it at the time, but my decision to work on a PhD was an unintended decision to not have more children. To me the most important job I have ever had is "Mom." My son Anthony was born in 1990 after I miscarried a baby in 1988. We definitely wanted more children. In 1993, my husband was offered a job in London working for the Air Force Office of Scientific Research. He said, "no, my wife has a career."

Enter two incredible men, then Colonel (eventually General) David Herrelko and Dr. Tom Curran. They came up with a solution. They offered me the chance to attend one of the top universities in the world to work on a PhD. While getting paid my full salary! To be honest, I was hesitant. My son was 3 years old and I had watched my husband earn his PhD—not easy. I turned the offer down twice. But I had this nagging feeling that I would regret saying no. Who in their right mind turns down such a "scholarship"? So, eventually, I said, "yes" (after swearing that only death would keep me from going to London). My husband had a housing allowance and we had a beautiful flat in Central London.

We were able to go "everywhere" at the weekends. But working on an engineering PhD, while your husband travels three out of four weeks of the month and you have a three-year old is *really* hard. My salary mostly went to a live-in nanny and a cleaner. I am very small to begin with, but I lost tons of weight. "Stressed" was an understatement. Pregnancy was not happening (which was probably a blessing). I did finish the PhD actually in record time. A couple of years after my husband passed away unexpectedly in 2016, I visited his London boss, Bill Crimmel, at his home in Colorado. I asked "Colonel C" what the plan was if Mark and I had not gone to London. He told me "there was no plan B, we all knew you would eventually say yes." It was indeed a "choiceless choice." But I did pay a personal price. By the time I finished the PhD and got back to a healthy weight, I was older and just did not get pregnant again. I do not regret choosing to go to London, it changed my life in amazing ways. But there is a part of me that wishes I had had more children. Luckily I have Santiago, my gorgeous Coton de Tulear puppy!

Mentorship/Sponsorship

How important was mentorship/sponsorship for your career? Have they been men or women? How was the relationship established?

Mentorship has been pivotal in my career. I entered the science and technology workforce at a time, 1983, when there were few women engineers. All of us have had "me too" stories when it comes to being mistaken for the secretary or propositioned by an unsavory character, among others. But I am a glass half-full kind of person. Being a woman also got me noticed more. Within the Air Force Research Laboratory,

I met a lot of amazing men who helped me along (at the time, there were not any women in positions of leadership!).

The most pivotal person was Don Daniel, who I met in 1997 after I returned to Dayton from my studies at London's Imperial College. Don was the Executive Director of the newly formed Air Force Research Laboratory, and he was a champion of women. He helped me get assignments that got me promoted. When he took a job at the Pentagon, he asked me to come work for him. I learned so much about the intersection of science and policy and how to deal with Congress! He was pivotal to my getting an assignment in 2001 to attend the Industrial College of the Armed Forces (ICAF—now the Eisenhower School for National Security and Resource Strategy). This was a school that promising future U.S. and allied military leaders are sent to, along with very few civilians like myself.

It was amazing (and scary) that I was part of what became known as "the class of 9/11." At ICAF I learned about grand strategies and honed my geopolitical skills and ability to negotiate in multicultural settings. Don gave me tons of opportunities. He was the toughest boss I ever had, one of his favorite phrases was "one oops wipes out a lot of atta girls." I also can honestly say that Don was a catalyst that transformed me. He saw me as a person that happened to be a woman, and he helped me believe that the executive ranks were inevitable for me. After ICAF, an executive job as Chief Scientist for Environment and Energy was available at the Federal Aviation Administration (FAA). The amazing Carl Burleson, the Executive Director, was looking for a scientist with policy and Congressional experience (quite an oddity!).

I was sent the job description by a few friends along with a "ha, ha, this was written for you." Ironically, Carl and I were stationed in London at the same time, but we only met when I interviewed. The mentorship and opportunities provided by Don Daniel made me "perfect" for the job, and Carl became my boss. This career vector was a capstone of all the experiences I had had up to that point. How did I get there? Amazing mentors! Also by turning a seeming disadvantage, being a woman in a male-dominated field, into one heck of an advantage!

Avoiding a Stall

How important is an advanced degree?

I think the answer to this question is "it depends." There is a quote by the Spanish writer, Antonio Machado, that I love. "Caminante no hay camino; se hace el camino al andar." Loosely translated (it is hard to capture the feeling of a romance language in English!), it means "Walker there is no road, you make the road as you walk.". I have two master's degrees and a PhD. But I feel that I earned them as part of the road I was walking, not as a much-thought-about "grand plan." When I worked at the Air Force Research Laboratory, advanced degrees were generally (but not always) necessary to get promoted. That motivated my first master's, Aerospace Engineering, somewhat. Also my husband wanted to pursue a master's, and it was something we did together.

I dramatically vowed I would never pursue a PhD when he went on for his; only to change my mind when living in London was tied to working on a PhD. So my PhD was driven by a desire to live in London! I earned a master's at ICAF, but again that was incidental to attending a senior military school. At the FAA, advanced degrees did not seem to be as much of a requirement. But the thing is you never know when that advanced degree will come in handy. There are some jobs that do require a PhD in Science or Engineering. A master's degree does differentiate you from other job candidates. Having a PhD does open doors. "Dr. Maurice" was somehow a lot more credible than "Ms. Maurice" at ICAO!

At the end of the day, my pursuit of advanced degrees was as much about the process as about the degree. For example, I feel that a PhD taught me how to think through a problem and how to persevere. It is a lot like childbirth. Both require breathing through pain and are so rewarding when you are finished! A PhD also teaches you humility and perspective. On the day you defend your thesis, you know more about your subject than anyone in the world. You also know that you barely know anything about it.

Plus, you have experiences and make lifelong friends and connections that help throughout your career. In my case, at Imperial, I learned how to read folks from many different countries, which was incredibly helpful in my work at ICAO. I have no "specific advice" for others; advanced degrees were helpful to me. But my sister was just as successful without one. Everyone should make their road as they walk! Learn, lead, serve. That is the motto of the University of Dayton, and one I try to live by.

Powering On

Have you ever cried at work out of frustration or anger? If yes, how did you handle that?

I am Latin American, so showing emotions is intrinsic to who I am. As a child in Havana, Cuba, I cried A LOT. I had a Spanish nanny who told me "save your tears for when you really need them." For some time, I felt she must have been right as work never caused me tears! But there was the intersection of the personal with work. My mom inspired my sister and I to pursue engineering, having everything you own suddenly "nationalized" by a Dictator (during the Cuban revolution of 1959) quickly teaches you that an education is the one thing no one can take away!

My mom passed away from cancer when I was 23 and I lost a pregnancy at 27. Those two experiences made work seem a lot less dramatic. I wanted to cry over those losses! But I didn't feel this was good for my career. The bathroom was a good place to go cry. Was that necessary? Maybe not. I guess I will never know.

I recall two times, when my career was already firmly established, that I cried at work. As FAA's Executive Director for Environment and Energy, I was the U.S. Member to the International Civil Aviation Organization Committee on Aviation Environmental Protection (ICAO/CAEP). CAEP negotiates environmental standards for international aviation. At the 2013 meeting of CAEP, I was forced to read a statement that was not data driven. The details do not matter; it was a lack of understanding rather than malice. The "deal" was I just had to read the statement and did not have to defend it. As I expected, every other CAEP member and Observer basically said what a daft idea the U .S. Member had just presented.

In many ways it was theater, but afterwards I was so embarrassed that I sat outside the room by myself crying while everyone else went to lunch. The French delegation came over to console me, which was very kind. I did not feel the need to "handle" the situation, I simply said "I am crying because I just committed intellectual prostitution." Perhaps a bit dramatic, but I take "data driven" policy decisions very seriously. I did cry over a work-related matter a second time when ICAO agreed on a global climate deal for international aviation in 2016. Most people were crying, it was so moving!

I feel that crying is no longer the taboo that I thought it was in the 1980s when I lost my mom and my baby. When my husband passed away suddenly in 2016, I often broke down crying in my office. My boss and my colleagues comforted me, which is the human thing to do! Tears are a healthy expression of all sorts of feelings, not at all a sign of weakness. And my nanny was not exactly right, I have yet to run out!

Beth Moses

Astronaut and Chief Astronaut Instructor
Virgin Galactic

About the Author

Beth Moses is an American astronaut, aerospace engineer, and chief astronaut instructor at Virgin Galactic, a public spaceflight corporation, which is developing suborbital space tourism. Beth leads Virgin Galactic's cabin test program and will train all SpaceShipTwo cabin astronauts for their missions. On February 22, 2019, Moses launched to space with pilots David Mackay and Michael Masucci and performed the first-ever test of SpaceShipTwo's customer cabin. During her test she became the first human to unstrap and float weightlessly in a stationary spacecraft, hovering completely still inside a spacecraft as it coasted to a stop above the Earth. Ms. Moses is also the world's first female commercial astronaut and the recipient of the Federal Aviation Administration's commercial astronaut wings #007, having been the seventh person in history to serve on the flight crew of a commercial spacecraft.

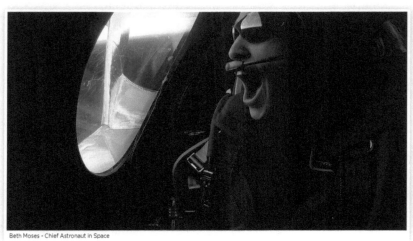

Beth Moses - Chief Astronaut in Space

Beth Moses

Previously, Ms. Moses worked at the National Aeronautics and Space Administration (NASA)'s Johnson Space Center where she served as the Extravehicular Activity System Manager for the International Space Station from design through on-orbit construction. Ms. Moses received her Bachelor's and Master's degrees in Aeronautical and Astronautical Engineering from Purdue University.

Beth Moses

Questions and Answers

Personal Career Insight

What was the coolest thing you experienced in your career?

Space! On February 22nd, 2019, I became the 571st human in history to reach space. I was there to conduct the first cabin test of Virgin Galactic's SpaceShipTwo. I later reflected on it with a colleague who asked the following:

When did you feel zero-gravity?
As soon as the rocket motor shut down, it was very smooth and the nose was pointing straight up. A few seconds later I was cleared to unbuckle. I put one hand on the window frame and with the other hand unclipped my belt and I was free. Hollywood movies always show people's arms and legs floating upwards as soon as they hit a zero-gravity environment, but in reality, there is an absence of any force. It just feels so natural. It's free and delightful and you can't help but smile.

What was the first thing you did in space?
Despite being terribly eager to release my seat belt and press my face to the window, the first item on my checklist was a planned safety check—I checked the condition of the cabin and my own condition to ensure that it was safe to unstrap and leave my seat.

As soon as I unbuckled, I turned to the window and saw the Earth below. My first thought was "I see the limb of the Earth!" The pilots told me afterward that I actually said it out loud. I was completely awestruck.

Then I continued with my planned test: I got back into my seat to check it, unbuckled again a different way, leaned and floated to the top of the cabin, translated purposely about checking specific handling aids and ship motion, and finally joined my pilots at the cockpit as we approached the highest point of our flight—apogee. The cabin felt just the right size as I was never out of reach for something to touch to help me move around.

What was apogee like?
Apogee was the high point in more than one way. It was the most magical moment of my life and there are no words to adequately express the feeling it instilled in me. The spaceship coasted to a complete stop and I was totally weightless, hovering in midair behind the pilots, miles and miles above the Earth. I became aware of the deep, dark eternal blackness of space, and then the Earth captivated my view. I could see a thousand miles, halfway up the U.S., halfway down into Mexico, the Pacific Ocean, and the North American continent.

During our mission, there was snow on the mountains that line the Pacific coast. Picture seeing snow in sunlight and then try to imagine it without the atmosphere. It felt like these mountains were glistening just for us. I will forever love the sight of a snow-capped mountain in a way I never did before.

Space is by far the most profound thing I have experienced in my career.

Work/Life Balance

Did you ever feel guilty for leaving work on time to attend a family event? Did you feel it reflected negatively on you at work?

No.

I tend to stay at my desk later than those around me, but of course, I have occasionally left work early. I've never felt guilty nor that it reflected poorly on me. I have been fortunate to work in organizations that respect work/life balance, and I've never abused the privilege.

However, even within a respectful workplace maintaining a balance between work and self can be challenging in the highly complex team environment of aerospace. Happily, the industry affords engineers a degree of personal flexibility as we aren't necessarily chained to the clock but rather responsible for a steady measure of progress. You can leave early to jet off to a concert but you'll typically be working on the plane and jetting right back. (Been there done that!) This mad-dash form of work/life balance isn't for everyone, and some forego the effort, in fact friends and colleagues find it a bit crazy. (They clearly lack my exquisite musical taste.) But I wholeheartedly recommend it whenever your project allows since there will be months and even years where you simply cannot break away.

I've now slogged through the development and verification of two human spacecraft, the International Space Station and Virgin Galactic's SpaceShipTwo, and in both cases, there were times when the core experts were working all night and communicating all day. None of us could stop and smell the roses or hear the music.

In that environment you can still leave work early for family events but in a way you don't ever really leave work. The hard truth about human spaceflight is that there are times when a balanced life cannot exist. Work is your life and you simply have to get the job done. I'd like to say you eat, sleep, and work, but the first two are optional.

Mentorship/Sponsorship

How important was mentorship/sponsorship for your career? Have they been men or women? How was the relationship established?

Very.

I have never been assigned a formal mentor or sponsor, but a science teacher at my high school, Ms. Lynne Zielinski, and an engineer at NASA, Mr. Fernando Ramos, were instrumental in my early years.

I was fortunate to attend a public high school that valued science and had a science club led by a physics teacher, Lynne. During my four years there Lynne became my informal advisor as we shared our love for all things space. Lynne was a finalist in the Teacher in Space program and helped me research the industry and evaluate universities. She brought enrichment programs to my attention and nominated me for substantial educational programs including the opportunity to attend a college astronomy course at Chicago's Adler Planetarium. Over the years we forged a friendship that I still treasure. I hope I would have found my way into aerospace even without Lynne's guidance—I was doggedly single minded—but it would have been a faltering journey. Lynne pointed out the road ahead at every turn and helped me stay firmly on it. I arrived at the start of my college years with confidence thanks to her unfailing enthusiasm and faith in my abilities.

A few years later I was in the midst of a student work program at NASA's Johnson Space Center when I found myself under the wing of a senior engineer, Mr. Fernando Ramos, aka Ferd. I have a hunch that one day his supervisor said "Ferd we have this student engineer who wants to help with the spacewalk testing. Please look after her for the summer. Good luck!" Fortunately, Ferd took it seriously and provided exceptional professional guidance. He always took the time to mentor me not just manage me. He would explain the test objective and share all the relevant technical background before assigning me a substantial portion of the work and letting me tackle it. He would unerringly steer me toward creating an exceptional test plan and solid test report that NASA could trust even after I returned to school. He also unfailingly gave me credit for my own work and made sure that his superiors, typically Shuttle astronauts, knew of my work and provided constructive criticism. He also kept me firmly on the critical path. I emerged from my student days a grounded and proven test engineer with genuine experience and a bit of street cred thanks to Ferd's patient tutelage. I am eternally grateful. I frequently think of Fernando when I work with young engineers—I can never repay him, but I pay it forward at every opportunity.

Thank you, Lynne and Fernando!

Avoiding a Stall

How did you develop organizational savvy? And how did it help your career?

Deliberately.

This question amuses me because long ago on the first day of the aforementioned student job at NASA I registered this stray remark: "NASA co-ops can move about the organization easily but permanent employees cannot." That shaped my NASA career, which led me to Virgin Galactic and ultimately to space.

I was part of my university's co-operative education program and slated to execute several semester-long engineering internships at NASA's Johnson Space Center before graduation. The comment struck me as important, and although I originally sought to work strictly in one office (Mars or Bust!), I decided to deliberately move about the center and experience various projects while moving was supported. Thankfully so because goodness knows I was a headstrong youth not otherwise inclined to slow down.

On the second day of my student job I was given a peculiar task: "Read these facility manuals and write a ten-page report on how mission control is reconfigured between flights. You have four months. Go." Seriously? A book report? I turned in my excellent ten pages (his words) that Friday and he had no other work for me.

Seriously?

Other students told me of a magical Weightless Environment Training Facility where engineers tested spacecraft mockups underwater. Now we're talking! I found it and presented my SCUBA certificate and spent the remainder of my semester designing, building, and installing underwater mockups for spacewalkers to establish the clearance needed to replace space station batteries. (Ever wonder why they are spaced as they are? That's why.) I was gleefully contributing to the agency's critical path.

From there I set about deliberately moving each semester despite having a killer job that wanted me to stay. I chose to work in four core technical areas—space/life sciences, engineering, mission operations, and flight crew operations—and capped it off with a summer at the launch site. I then settled into my chosen area, spacewalk verification, before finishing graduate school and accepting NASA's coveted job offer to join flight crew operations as a support engineer.

I didn't enjoy uprooting myself so often. In truth I hated leaving each group and starting over every semester but it was my only chance to do so easily and it allowed me to apprentice across disciplines. I learned how to see projects from various angles. This would later enable me to contribute meaningfully to ever larger and more complex efforts. More importantly, this approach endowed me with the technical chops to integrate complex crewed engineering projects, which is the backbone of human spaceflight.

From there it wasn't a matter of avoiding a stall but rather keeping up since I was continually in demand. I very happily grew from a teenage mockup engineer to a test program lead to the International Space Station (ISS) spacewalk system manager. When ISS assembly was complete I avoided the siren song of sustaining engineering by leaving NASA to sink my teeth into commercial space another self-inflicted uprooting to learn new ropes.

I guess a good way to prevent a stall is to build yourself into the best, most technically-savvy workhorse you can and then to hang on for dear life as your career unfolds. Deep yet broad expertise is a formidable career accelerator.

Powering On

Was there a significant event that changed your career trajectory and what was it?

Yes. My spaceflight in February 2019 to conduct Virgin Galactic's first cabin test aboard Virgin Space Ship (VSS) Unity.

The test was profoundly important for Virgin Galactic. The company was formed to fly humans in the commercial cabin of SpaceShipTwo, and before early 2019 we had not yet achieved this goal although we had sent an instrumented dummy to space a few weeks earlier, so I like to joke that we replaced one dummy with another-Myself! Joking aside, I felt acutely responsible to conduct the singular most efficient test possible since the entire team's effort to date was culminating in my cabin test, which would establish our corporate footing going forward.

So I constructed and optimized safe test in three phases: a ground run of the customer training program with the full ground training team to rehearse and refine it for customers, a series of in-flight observations from my seat to roll into customer training plans, and detailed weightless procedures to put the cabin through its paces and seek out the unknown unknowns. I then constructed, practiced, and memorized three alternate sets of weightless procedures in case my weightless time happened to vary from our preflight predictions.

On February 22, 2019, test pilots Dave Mackay and Michael "Sooch" Masucci flew VSS Unity to space so perfectly that I was indeed able to leave my seat and conduct my test safely. I recorded hundreds of observations and ratings alongside relevant sensor data.

After we landed safely back on Earth, I knew we had a product and I came skipping off the ship with a huge smile as a walking, talking proof of concept.

This cemented our program credibility and helped to bring us secure crucial investments. We took the company public, relocated to our commercial spaceport, and ramped up toward commercial service. It also cemented my role as the first, most expert SpaceShipTwo cabin operator and demonstrated my proficiency during a demanding test flight. I emerged eminently qualified for the remainder of my duties: leading the rest of our cabin testing, training other staff to fly, and training our customers.

It also happened to anoint me the world's first female commercial astronaut but I never set out to be historic or heroic, I simply wanted the job of astronaut and wanted to do it well. I'm humbled that my work has been noticed outside Virgin Galactic, and I was honored to be asked to contribute my thoughts to this publication alongside my industry sisters. I hope my words have helped you, the reader, in some small way. I hope it gives you proof that the impossible is not actually impossible and shows that aerospace is tough but women are tougher. And I hope that someday you get a chance to see the Earth and space as I have: unstrapped and hovering weightless in a completely still spacecraft gazing down at the magic of our home planet.

Beth Moses

Beth Moses

Doline Peterson

Program Manager
Safran Power Units

About the Author

Doline Peterson's story is a journey through drastic weather that led her to sunny San Diego, CA. She was born and raised in Cameroon, in a region where the weather hardly ever gets below 70 degrees. She earned her undergraduate degree in Physics, with a minor in Computer Science. The school Doline attended, the University of Buea, did not offer major programs in computer science at the time, but she always wanted to work with computers. After her degree, she worked in the city of Douala, Cameroon, for three years as a computer networks engineer for a major network service provider. She designed, installed, and maintained computer networks, Voice over Internet Protocol (IP), and IP Telephony solutions to connect nationwide businesses with their worldwide branches. While on the job, she earned coveted Cisco certifications, but what she really wanted was to further her education.

Doline moved to Grand Forks, ND, in the winter of 2006, to pursue graduate studies in computer science. She went from 100-degree weather to minus 20 in a couple of days, all only wearing a cute sweater and a scarf, the only way she knew to deal with cold days. She had to adapt quickly, just like for the rest of her journey. In Grand Forks, she was introduced to Aerospace. As a graduate research assistant, she worked with the Center for People and the Environment on the Agricultural Camera (AgCam) Project, a National Aeronautics and Space Administration (NASA)-funded program that sent two cameras to the International Space Station in 2008. She found the AgCam project to be a very fulfilling experience and considers it the coolest thing she has done in her career. In March of 2008, Doline was invited to the Kennedy Space Center for the payload launch onboard the Endeavor space shuttle mission

STS-123. She had the chance to visit the launch pad with the raised shuttle the day before the launch and was invited to view the launch from the KSC launch press site.

Doline's excitement for AgCam was renewed this year when Robert Behnken, a NASA crewmember from the space shuttle mission that delivered AgCam to the ISS, named SpaceX's first crewed capsule Endeavor, after the retired shuttle.

After Grand Forks, Doline spent the following 12 years of her career in systems and software engineering for business and regional jets. Her main assignments were the development, verification, and certification of Flight Control Systems, Flight Management Systems, and Display Systems with Rockwell Collins, then of Auxiliary Power Units (APU) with Pratt and Whitney, and now with Safran.

Doline, her husband, and their daughter really enjoy all things Aerospace. They went to see the Perseverance Rover at the Jet Propulsion Laboratory before its launch on its mission to Mars and signed up to add their names to the rover. They went to see the first launch with a first stage returning to land in California at the Vandenberg Air Force Base. They also saw the retired Endeavor shuttle, the first landed orbital class rocket by SpaceX, and the James Webb Space Telescope.

About a year ago, Doline was offered the opportunity to join her business unit's leadership team as a program manager, which is her current role. She oversees the development, certification, and production of the APU on Dassault's F6X business jet. She leads an Integrated Program Team of leads and managers that comprises engineering, finance, supply chain, procurement, production, quality, customer support, and aftermarket. Her primary responsibilities are focused on coordinating initiatives across these disciplines toward two main objectives: customer satisfaction (upholding Safran's end of the contract with Dassault) including engineering and production deliverables, and toward program profitability (meeting the costs and revenue targets in the program's business case). She is learning a lot from all aspects of the business, from all disciplines, and on how synergies between these disciplines work to meet the program's objectives and affect her business unit's Profit and Loss statement.

Doline hopes that her journey through drastic weather is over, but that her journey outside of engineering is just beginning, and she looks forward to expanding her knowledge.

Questions and Answers

Personal Career Insight

How did you decide between a leadership vs. technical career track?

Up to a year ago, I was well engaged on a technical career track. I built relations with the standards organization governing my domain, the RTCA, and served as the primary representative for my business unit. I was introduced to the Federal Aviation Administration (FAA) as the business unit's Subject Matter Expert (SME) in training, and with the current SME, developed a plan to obtain the Technical Standard Order SME approval by the FAA, which would grant delegation of authority to conduct audits.

Doline Peterson

Doline Peterson

Throughout my career, however, with increased responsibilities, came the need to acquire project management skills, then the need to understand Earned Value Management, and then the need to develop a basic understanding of financial principles, all while leading engineering efforts toward systems certification. To acquire these skills, I completed a Project Management Professional certification and then enrolled in an MBA program with San Diego State University.

These decisions created an opportunity for me in Program Management that I did not anticipate. An opportunity I was excited to explore. I am currently managing a program that is in its development phase, and having a solid engineering background is definitely a plus. Meanwhile, I also get to learn about other aspects of the program, such as establishing a supply chain, developing manufacturing processes, building aftermarket strategies for entry into service, and the process of defining business strategy.

Work/Life Balance

Did you ever feel guilty for leaving work on time to attend a family event? Did you feel it reflected negatively on you at work? If so, how did you deal with that?

My priorities have always been clear: health, family, and then my career. This clarity or order does not absolve me from putting in the work necessary to advance my career. However, should there be conflicts, my choices and decisions are mostly predetermined. The toughest decision I had to make in that regard was to quit my job to reunite my family.

After Grand Forks, I moved to Cedar Rapids, IA, with much warmer weather, where I worked for Rockwell Collins for a little over four years. I had started building a family that grew with the birth of our daughter. My husband and I had stable jobs, but his was in California. I first looked for opportunities to move within Rockwell Collins, but the industry was slowly recovering from a recession, and that limited my options for transfer. This led me to leave Rockwell Collins to pursue a job with Pratt and Whitney in California.

I have been fortunate to work with companies that value family. In my current job, my Vice President and General Manager has made it abundantly clear that family is the unequivocal number one priority. We all work extremely hard with a very lean staff, but no one feels constrained to choose between family and work, and we all feel comfortable putting family first when a conflict occurs. This makes the work environment so much better, and I really enjoy this dynamic.

Mentorship/Sponsorship

How important was mentorship/sponsorship for your career? Have they been men or women? How was the relationship established?

Mentorship has been defining in my career. My first mentoring relationship started at Rockwell Collins, through a company-wide program. The program was promoted via emails, was recommended by a friend, and caught my interest. I was curious to find out what could come from it. I was not prepared for how eye-opening casual lunches, with someone who has been where you are, could be. I learned a lot about myself, my strengths and weaknesses, and what the world had to offer. She pointed out interesting things about me that I would have never otherwise valued. She opened my eyes to options that I would have never otherwise considered, explained how my department worked, and described the career paths.

Since then, I have made sure to build mentoring relationships, both formal and informal, and I have had more women mentors than men, all outstanding and inspiring individuals. For many, the mentoring relationship was established by me just asking, no predefined formats, just building a relationship.

I am deeply moved by the selfless individuals who give back and mentor, often through professional societies, such as the National Society of Black Engineers (NSBE) and the Society of Women Engineers (SWE), and who build a network to support and sponsor early-career engineers. I always make sure to let interns and new hires know that I am available as a mentor.

Avoiding a Stall

Were you ever presented an "opportunity" you declined and did it hurt your career? If not, how did you overcome any negative impact?

Envisioning where you want to be in five, then in ten years, helps identify the opportunities you want to prepare for and gives you the confidence to turn down the opportunities you do not want, no matter how attractive they are.

I have had several examples, throughout my career, of turning down opportunities sought after by many of my peers because they did not align with my longer-term aspirations, including when I decided to quit my job to reunite my family.

One opportunity that stands out to me is when I was offered to take a role with a newly formed joint venture between Safran and Boeing. Opportunities to work with

Original Equipment Manufacturers (OEMs) are typically much more sought after, than working with the lower-tier suppliers.

Although I was very grateful for the opportunity, I made the decision not to pursue the path that was opened to me, because the specific opportunity didn't align with my plans for growth. Either on the technical or management tracks. Later on, I was offered another opportunity that I was really excited about, and I consider myself lucky to have no regrets. There is always some uncertainty when turning down an opportunity to assume a different role, or an opportunity to earn more money by reconsidering your plans for a change and remaining in your current role.

Powering On

Have you ever cried at work out of frustration or anger? If yes, how did you handle that?

I have never cried at work out of frustration, not because there were no frustrating situations or causes to cry, but because I have seen my mentor cry out of frustration from a meeting. I am grateful for that moment. Seeing vulnerability in strength, cemented my understanding that, along with the power dynamics, a cross-functional leadership board also has its interpersonal dynamics. Encounters and exchanges are not all going to be smooth and rosy.

I can recall a situation where this understanding came in handy. In a meeting, I interrupted a peer to express my disagreement with his plan, things we all do occasionally, most times without realizing it. Because of the sensitivity of the issue at stake and its importance to the business, my coworker lashed out in anger for being interrupted. He then restated his plan and changed it to be more in line with my recommendation. His reaction created quite a memorable scene, but as I had apologized for the abrupt interruption, I made nothing of it.

After the meeting, I received a call from our common supervisor, who was also part of the meeting and wanted to make sure he gave me a piece of his mind. I felt that I was being yelled at unfairly. However, because I remembered my mentor, I found the strength to not respond or pick up a fight I'd lose, and just let it go and apologized, again. To my surprise, my reaction turned things around and earned me a shower of compliments at my next performance review. I was praised for my handling of stressful and contentious situations, praises that ended with "I trust you." Apology accepted and you are welcome! I didn't say those last words out loud, of course. This supervisor turned out to be a mentor I trust and look up to.

My take away was that you should allow yourself to be vulnerable, and most importantly, forgive yourself. Don't let past mistakes hold you down. I did not have this insight early in my career, and I tended to punish myself.

Deb Rayburn

Customer Support Director
Business and General Aviation
GE Aviation

About the Author

Deb is the Business and General Aviation Customer Support Director at GE. Aviation based in Cincinnati, Ohio. In this role, she leads a global team that supports business and general aviation jet engine operators flying around the world.

After working at GM as a Manufacturing Engineering for four years, Deb joined GE Aviation in 1989 as an engineer working on design and external technical integration efforts for commercial engines. In 1997, she became a Field Service Engineer working in Seattle and managing airframer and customer efforts on multiple aircraft models.

Three years later, she returned to GE Aviation's Cincinnati headquarters and worked in Customer and Product Support where she grew her global customer and airline operations expertise. In 2004, Deb assumed New Product Introduction and engine certification responsibilities as a Senior Staff Engineer in Systems Engineering. She subsequently worked on the GE Honda HF120 program as the Technical Program Manager at Unison Industries, a wholly owned subsidiary of GE Aviation, and as Flight Test Integration Leader.

In 2014, Deb became the Commercial Engines Field Service Manager, leading a team supporting Delta Air Lines. During the next four years, her role expanded to cover all the regional engine operators in the United States.

Deb joined the Regional Engines and Services team in 2018 as the Product Operations Manager, coordinating with Supply Chain Management to improve CF34 engine deliveries to Embraer. In 2019, she took on her current role in Business and General Aviation.

She has a Bachelor's degree in Engineering from Purdue University.

Questions and Answers

Personal Career Insight

What was the coolest thing you experienced in your career?

After spending eight years building my foundational skills in design engineering and systems integration at GE Aviation, I became a Field Service Engineer at our customer Boeing and worked the wide-body flight line supporting 747, 767, and 777 aircraft deliveries in Everett, Washington. I realized that I wanted to be a field service engineer working with our customers within my first year at GE after I had the opportunity as a design engineer to work alongside the field service engineers who were supporting the Airbus production system. It was very formative to me to see these Field Service Engineers who got to spend every day with our products, our airframers, and our end customers as they took delivery of their airplane.

In this role, my days were spent working with Boeing technicians, engineers, and leadership as we prepared airplanes for delivery to airlines and leasing companies. My team of technicians and I worked with Boeing engineers and maintenance personnel to resolve technical issues with the engines during the engine installation and aircraft delivery process. The impact I made in that role and the relationships I built with Boeing and our end customers were exactly what I loved to do. In one day, I would work with a flight line engine mechanic and then later be talking with a Vice President of an airline or leasing company, answering questions about our products. It also allowed me to touch the global customer base and learn how to work with different cultures, which, as a female, can be challenging when dealing with patriarchal regions of the world.

One of the coolest things that I did in this role was fly in the flight deck on a 777 test flight. An engine had to be changed on an aircraft and someone needed to fly with the pilots for the reflight, which happened to be on Easter Sunday. I jumped on the opportunity and became part of the flight crew, wearing the headset, running the maintenance data computer, and reviewing fault data. As a non-pilot, this was a first for me as I had never experienced getting to be part of a flight crew on any flight. It made up for missing Easter with my family and friends.

Work/Life Balance

Did you ever feel guilty for leaving work on time to attend a family event? Did you feel it reflected negatively on you at work? If so, how did you deal with that?

For the first decade at GE, I gave everything to my work and career. I made a lot of personal sacrifices, and I did a lot of traveling. In the second decade, I made choices about family.

Deb Rayburn

After I returned to GE Aviation's headquarters in Cincinnati from Seattle, I met my husband Brian, and we married in 2003. I decided to slow down and try to start a family. This was a turning point in my life. During my time at GE, I never slowed down, and in my Boeing field service engineer role, I worked seven days a week, 360 days a year. But in the early 2000s, I realized my personal life was more important than my job. When I finally slowed down, I questioned my work/life balance, which was really out of whack. I had zero balance in my life when I was young. I started striving for that in the early part of the decade.

I was working as a Customer Service Manager in the Customer and Product Support organization, which was a role that I wanted for many years. I felt like my career was heading in the right direction. While my mom had been ill on and off for several years, my dad became very ill and was diagnosed with cancer in 2004. The doctors gave him four to seven years to live. Unfortunately, my family was very familiar with cancer diagnosis after my niece was diagnosed with leukemia in 1996. With my mom and dad both being ill, I was getting multiple calls throughout the day, and travel was impossible, which was a job requirement, so it was growing difficult to perform the role. I spoke to the leader of the Customer and Product Support organization about my situation who supported my decision to ask for a new role. I knew I could do this role, but I didn't have the capacity at that time to do it. I felt I was going to fail, and I couldn't let me, or GE, fail the customer.

My leadership helped me to transition back to Engineering and work engine certification and flight test integration roles so I could stay and stop traveling, yet work in an externally facing product-oriented role. In 2006, my mom's health took a turn and she spent time in the ICU, which included a lengthy recovery. My leadership gave me the flexibility to work non-traditional hours remotely so I could help care for her during the day and work when she was sleeping. My dad's immune system was down due to his cancer treatment, so he spent time in and out of the hospital during this time battling pneumonia. My only sibling—my brother—was getting his career back on track after taking time off to care for his daughter. So it was my turn to ease back on my career.

When I look back on that time from 2004 to 2010, I would not change any decision I made. The memories of those years will stay with me forever as my mom passed from pancreatic cancer in 2010. Fortunately, my dad has outlived his doctor's expectations and is alive and doing well today. Did I limit myself and my career with the choices I made? Absolutely, but I have no regrets.

Mentorship/Sponsorship

If you had a Sponsor to reach executive levels, how did you meet that person? Was the person assigned to you or was it a boss that became your sponsor? How did this person help your trajectory?

I was fortunate to have several senior-level executive sponsors during my career. I met my first sponsor through GE Elfun, which is an employee-led philanthropic organization. He was president of the Elfun Board, and I was a board member. The Elfun group was celebrating its 75th anniversary of community engagement. The Board held a Drive for 75, which meant we were pushing to do 75 community volunteer projects in Cincinnati, Ohio (our HQ location) in 75 days. We blew it away, and the team completed 104 projects in 75 days. He appreciated my energy, enthusiasm, and drive, and he recruited me to join his organization, Customer and Product Support. He encouraged me to find my voice, take a seat at the big table, and be heard in meetings.

I was working in his team when my dad was diagnosed with cancer and I had a miscarriage all within a four-month period. It was one of the worse years of my life, and I knew I couldn't perform in this key role, which I had worked to achieve for 15 years. I was angry, brokenhearted, and scared. While my direct manager wasn't supportive, his manager who was my sponsor, was completely supportive. He gave me permission to jump off my career track. He told me some things are more important than your job and taking time out is okay. He said the company would work with me and get me back to a better place that worked with the constraints of my life. Then he wrote me a note that said "Your manager cares about you. Your manager supports you. You're going to be okay. You need to do what's best for you." I carried it around with me for a long time. No matter how bad things got, I would look at it and I knew I was going to make it.

I have another strong senior advocate who I had met earlier in my career at GE. He helped me relaunch my career when I was ready in 2013 and has been my advocate and supporter for the last seven years including achieving my latest position He hired me into the Commercial Field Service organization where I had the opportunity to relocate to the field and lead a team of field service engineers located at ten different customer locations. He let me grow my team with the right talent, had trust in me to handle difficult personnel decisions, and always valued my opinion. When the time was right to return to HQ, in 2018 he rehired me in a role within his new organization, which allowed me to continue to grow and broaden my knowledge in an area where I had little prior experience. He's given me permission to call him any time for advice or to just talk. He allows me to be my authentic self and has enabled me to bring my whole self to work, which hasn't always been true throughout my career.

Avoiding a Stall

With respect to your career, did you ever hit an organizational roadblock? How did you overcome it?

By 2013 after my grieving for my mom subsided, I felt I had more to give and wanted to take that next step and become a people leader. I felt I had a great career here, had much to give back, and my experience could benefit others. Like all companies, I had worked for good and bad leaders. I knew what it took to be a good leader, mentor, and coach for future generations. I approached my leadership about moving into a people leader role. But they were not encouraging.

Yet I didn't want to give up on what I felt was the right move. So I networked and talked to a co-worker who I'd known for several years. We talked about my favorite role, which was my field service role and my passion for working with customers, our engines, and airplanes. At that time, the field service team was looking for women to join the team as leaders, and she thought I'd be a perfect candidate. They needed a new field service engineering leader based at Delta Air Lines in Atlanta, Georgia. My husband supported my move to this new role, so we made the move to GA. We were fortunate that he was able to transition his role to a remote role, which simplified our decision. Over the next few years, my team expanded to include all the regional engine operators in the US, with the role of becoming one of the most fulfilling and rewarding ones of my career as I discovered that I thoroughly enjoy being a people leader and developing talent.

Powering On

Have you ever cried at work out of frustration or anger? If yes, how did you handle that?

Absolutely. I don't yell when I get angry. Instead when I get very angry and frustrated, my emotions rise, and I do cry. Crying is my outlet. I've learned through the years that tears give me strength. After I cry, I get clarity. It's a release mechanism for me. It stops the world from spinning and shows me the right path to follow.

I've learned that bottling up your feelings…stress, anger, frustration … is like a pressure cooker. If you don't let off steam intermittently, it will explode. I've learned how to deal with my emotions over time as I've learned about self-regulation. I've recognized that being candid and transparent is what prevents my frustration level from building and the tears from flowing. Learning these skills, plus being humble and authentic, has helped me to grow as a leader and become much more effective at work. I can bring myself to work and deal with situations in a more constructive and productive manner and leave work without feeling emotionally drained at the end of the day.

Most male managers don't handle tears very well. They view it as a weakness instead of an emotion to be dealt with. Tears aren't a bad thing. Tears make you real. A few years ago, a male employee who worked for me was upset after not getting a new role. We talked, and he shed tears as he talked about his daughter's struggles and why he wanted to find a new role in a new city to help her. He wasn't ashamed of showing his emotional side. I told him he'd find the right role, that there was something else out there for him, and that I fully supported him as he had the talent. A better role surfaced within the next six months that was the right fit for him and his family and gave him the growth opportunity that he was seeking. He's in this role now and he and his family are thriving. I appreciated his authenticity and openness, which I find refreshing.

It's okay to cry but figure out why and develop techniques that help alleviate the stress that is causing the emotions. We put a lot of energy into our work. You spend more time at work than with your family, and you will get emotional at times… don't apologize but look to find ways to address what's triggering you before things get to this level for your overall wellness.

Tammy Reeve

Chief Executive Officer
Patmos Engineering Services, Inc.
Airworthiness Certification Services, LLC

About the Author

Tammy Reeve is Chief Executive Officer (CEO) of two companies: Patmos Engineering Services, Inc. and Airworthiness Certification Services LLC (women-owned small business). Patmos employs five full-time personnel and up to eight contractors and provides full design services for hardware and software for safety-critical and commercial systems. Tammy is an FAA-independent Designated Engineering Representative (DER) with Patmos Engineering Services, Inc.

Tammy has been working in the aviation field for over 30 years. Her work experience includes writing hardware test drivers and Programmable Logic Device (PLD) design at Unisys Corporation and embedded software design for GE Aerospace and Avtech Corporation. She has worked on aviation equipment ranging from flight controls to audio control systems. Tammy has a Bachelor's in Electrical Engineering (Tau Beta Pi) from California State University Fullerton (with emphasis in digital hardware design) and a Master's in Software Engineering from Seattle University. Recent areas of DER work include participation in the DO-254 workshop; SC-205 (DO-178C model-based subgroup as co-secretary), SC-190 working groups, reusable tool qualification, and development and training for the FAA Academy Airborne Electronic Hardware (DO-254) training course.

Tammy is a member of the RTCA Forum for Aeronautical Software and Chair for USA DO-254 Users group. Tammy is authorized as a Software and Airborne Electronic Hardware (AEH) (PLDS, FPGAs, ASIC) DER for levels A-E, and is also authorized for EASA validation programs. She is authorized for CFR 23, 25, 27, 29, and 33 as well as working many TSOs/ETSOs. Her interests include her family (including two adult children), friends, gardening, boating, and discovering new places.

Tammy Reeve

Questions and Answers

Personal Career Insight

If you could give early-career aerospace women one piece of advice, what would it be?

Look for places to be involved that are struggling, put yourself in places where you can be the person who is not afraid to dig in when things get tough. Remember two important things: "Risk = Reward" and " you don't get what you don't ask for." All they can do is say no, but if you don't ask, you will never know. I was always looking for places to support the project where it was struggling, to be part of a recovery path or success where there were issues, and I looked for opportunities to take on responsibility and stretch myself.

Work/Life Balance

Did you ever have to make a move to advance your career (within your company or changing companies) that impacted your family life, and how did you balance the two?

When my husband and I were ready to start a family, my company of several years was laying people off, so in interviewing for new employers, I was thinking and

Tammy Reeve

looking for companies that would support my goals to be both a mom and an engineer. Sometimes the smaller companies can be better suited to be flexible, and I ended up working for a small company, becoming an FAA DER and finishing my Master's in Software Engineering while having my two children (who are both engineers also! Proud mom!).

It took me five years to get through my master's, but I did it. The small company I worked for then supported me through it all and allowed me to work 20 hours a week for over 5 years of that time, which turned out to be a benefit for both me and the company at the time. Once my children were older and in school, I decided to take a big jump to start my own company Patmos Engineering Services, Inc., and now it is 20 years old. Being my own boss allowed me the flexibility I wanted to be involved with my kids, but pursue my professional goals and growth. Of course, none of this would have been possible without my supportive husband of over 35 years (who is also an engineer).

Mentorship/Sponsorship

What activities have you engaged in that have helped other women achieve success in their aerospace careers?

I personally mentored young women early in my career through SWE and presented at high schools on career day. I also hired student engineers as interns at Patmos to get experience over the summer. I review resumes and make referrals for young women engineers that I know and work with for job placement and career advice. I am proud to say that my daughter has her Master's in Material Science Engineering, I would like to think I had a hand in that.

Avoiding a Stall

How important is an advanced degree?

Pursuing an advanced degree for women is important. Unfortunately, it is still a male-dominated field, and I have found that women are continually underestimated. I think an advanced degree can help to combat this unfortunate double standard for women headed into a leadership or management role in engineering companies.

Advanced degrees should be chosen to enhance your career path, so once there is some idea what the path is, then go invest the time and money.

Powering On

Was there a significant event that changed your career trajectory and what was it?

I have been told from the time I announced I wanted to be an electrical engineer and was admitted to the University of Washington pre-engineering program, "you won't finish" or "you will quit" or "just change to something easier." All of these comments just solidified my resolve. When I applied to the University of Washington for the electrical engineering program in 1983, the program was really small with limited spots and very impacted.

It was not uncommon for only those with 4.0 GPA in the two-year pre-engineering classes to be admitted. I was told by the dean that if I was a minority, I might have a chance at getting in at that time. Many of the other female friends I had made decided to change the educational career path. But I was not going to let the university deter me from what I wanted to do. I had a 3.4 GPA and decided I would apply to other EE programs until I could achieve my goals. California State University, Fullerton (CSF) accepted me, and I became part of the charter group to start the Tau Beta Pi Engineering Honor Society at CSF. I graduated on the Dean's List.

Tamaira Ross

Director
New Glenn System Definition and Design
Blue Origin

About the Author

Tamaira Ross is Director of New Glenn System Definition and Design with Blue Origin, a Seattle-based aerospace manufacturer and suborbital spaceflight services company. She has the overall responsibility for the New Glenn system configuration and top-level interfaces as part of the systems architecture team.

Prior to Blue Origin, Ross worked for The Boeing Company where she led preliminary vehicle design and rapid development of aircraft and spacecraft programs. She was named a Boeing Associate Technical Fellow in 2009 and a Technical Fellow in 2014. Fellows are leaders in their technical disciplines, undergoing a rigorous selection process, representing less than 0.5% of Boeing's engineering and scientific workforce.

Tamaira holds a BS and an MS in Aeronautical and Astronautical Engineering from Purdue University. She also attended the University of Washington where she obtained an MS in Mechanical Engineering and a Technology Management MBA. In addition, she is an affiliate instructor in the Industrial and Systems Engineering department at the University of Washington where she teaches a graduate class in technical leadership. She is a member of the Purdue School of Aeronautics and Astronautics Industrial Advisory Council, the College of Engineering's Cislunar Initiative Advisory Board, and is a recipient of the school's Outstanding Aerospace Engineer Award. She is a Fellow of the Society of Women Engineers.

Ross has played golf for over 35 years, many of those competitively, and has learned several martial arts over 20 years. She and her husband enjoy exploring new places through hiking, road cycling, and mountain biking.

Tamaira Ross

Questions and Answers

Personal Career Insight

What inspired you to choose a career in Aerospace and what has given you the most satisfaction during your career?

I have always been fascinated by space travel, flying, and building new things. Growing up, I spent countless hours researching the solar system and reading about how spacecraft and airplanes worked. I dreamed of becoming an astronaut. Every LEGO set I owned eventually turned into spaceships at some point, rather than what the kit actually was designed for. As I got older, I learned that many astronauts and the people who built the vehicles they flew in were aeronautical engineers and that is what I set my sights on doing.

I had very supportive parents who encouraged me to work hard. The word "can't" was outlawed in my house growing up. My attitude was, and still is, anything you put your mind to is possible. I needed that attitude many times as the path was not always easy. I struggled in math in junior high and first years of high school. Fortunately, I was able to get additional help from other instructors who had different ways of explaining concepts. Looking back, I came to understand that it was not that I was incapable of doing math, it was that I needed a different way of learning. I've used this realization many times in my education and career. If I do not understand something, I seek out a different explanation rather than assuming I will not be able to understand.

I've basically had my dream job for many years. I have been very fortunate to go from building LEGO spaceships to building real spacecraft. I think that's a rare opportunity, and I'm extremely grateful to have the set of circumstances, work ethic, and skills that led to what I am doing today. I have worked on designing commercial airplanes, military airplanes, satellites, and now rockets. These are not easy projects, and work can be hard sometimes. Working on big, complex programs that involve a lot of people and tight timelines can be difficult, but in the end, I think it's worth it. I am most fulfilled when I see systems which I have designed, and have only been a vision in my mind for so long, come into reality by being built and flying. When projects are difficult, I have to look at the overall scope and breadth of the work that my team and I are doing and realize how important it is to our country and humanity. That is what drives me.

Work/Life Balance

Did you ever have to make a move to advance your career (within your company or changing companies) that impacted your family life, and how did you balance the two?

For me, this topic is about a move I did not make. I had the opportunity to move to Los Angeles and that move may have advanced my career. I was asked many times to make this move over several years. I chose not to do this because it was incompatible with my husband's career path. We are both engineers, and while engineers can work in a variety of locations and with many companies, it is not an entirely portable career when established. Building up a career in an industry takes time, and for many industries, there are only so many places in the country where that work happens. He would have had very limited choices in his industry in Los Angeles; therefore, we decided not to make the move.

Likewise, when he was offered an opportunity to move and to stay with his company at the time, he also declined that offer. I would have had very limited aerospace opportunities in that new city.

I would not say that these were necessarily hard choices to make. My husband and I consider ourselves members of a team, and like any team, success depends on each team member being able to fulfill their full potential and be happy. Both partners have to look out for their teammate when it comes to career choices.

Mentorship/Sponsorship

How important was mentorship/sponsorship for your career? Have they been men or women? How was the relationship established?

Both mentorship and sponsorship have been very important to my career, and I have been fortunate to have several great mentors. I have developed mentoring relationships through both formal programs and through informal means. Most of my mentors and sponsors have been men, although I have also had several women mentors.

Several of my mentoring relationships came about because I volunteered to do something outside of my regular work assignment like teach a class, connect with other engineers off-hours doing similar work to me, or get involved with professional organizations like SWE, AIAA, or SAE. I never have simply asked a potential mentor, "can you give me career advice?" as the basis for starting a mentoring relationship. Mentoring relationships take time to build and asking this type of question takes time for the potential mentor to know you in order to give a good answer. Mentoring should go beyond the model of meeting periodically and only talking about career plans. There are many ways to mentor and to be mentored such as doing a project together or teaching a class together. These experiences tend to be richer in content and learning. Through that process, the mentor ends up giving career advice along the way.

In terms of formal programs, I have been in leadership development programs in which participants chose mentors. I approached this with a specific goal in mind. At the time, I was trying to decide between going into management or a technical career path. I chose a mentor from each path, one who was a woman in senior management and a man who was a Technical Fellow. I met with both of them, observed and learned from them, and did projects outside my normal work with them. At the end of that process, I was more confident in choosing the Technical Fellowship path as a good fit for me.

Knowing what you want to learn from a mentoring relationship is important. Everyone has something to teach. Observe people who have skills you would like to learn. Ask yourself, "what makes them successful in that area and can I learn it?" If so, ask that person to help you learn that skill. This targeted approach can help you pick up new skills from many people. Keep in mind that mentors do not always have to be older than you are. Especially with a targeted skill approach, they may be younger than you are. You also may have several mentors at one time.

Sponsorship is different than mentorship and it is important not to think the two are the same. Sponsors have the ability to advance your career through their position and influence. Mentors will help you get to the point where your career can be advanced. I have had mentors become sponsors, but this is not always the case. Sponsors will advocate for you even when you are not present, making sure you are considered for promotions or more challenging assignments. Mentors can prepare you for these roles, but they typically will not make them happen.

I have also served as a mentor to many people, especially women. I have organized programs for university students to be mentored by professionals and mentored many students myself. I have also had years-long mentoring relationships with several protégés. I have learned just as much from mentoring others as I have by being mentored. Mentoring provides an opportunity for me to reflect on and learn from my own experiences by relaying them to another person. In addition to helping others, I'm also solidifying my own experiences and learning from them.

Avoiding a Stall

How important is an advanced degree?

I have three advanced degrees, so I suppose it is somewhat self-evident that I think they are important. However, I did not do any of them simply for a credential or as a resume builder. They were all too difficult to achieve if that were the only objective! I believe in pursuing advanced degrees for yourself and your own learning. I do think advanced degrees can advance a career, but it is because they can change the way you think, which can then be applied towards more difficult and different challenges over the course of your career.

My first MS degree from Purdue University in Aeronautics and Astronautics, with a thesis, taught me how to solve a novel problem independently. That is the primary objective in doing a research-based MS. I also learned how to work with professors as research advisors rather than only teachers I took classes from in under-graduate work. This was important as I continue to interact with these professors professionally today.

My Purdue education gave me a great foundation. I've done a lot of different things in my career and worked on a lot of different programs and vehicles. Purdue taught me how to learn, and that was the biggest thing I got out of my education. It taught me how to solve problems that aren't in a book. Ironically, formal education spends a lot of time using books, and I really learned the fundamentals, but what my MS degree really taught me was how to teach myself throughout my career. That ability has proved to be critical and has made me successful working on projects where everything is new and almost nothing can be found in a book.

My second MS degree from the University of Washington in Mechanical Engineering taught me to branch out into new fields. When I joined Boeing Defense, Space, and Security, I was encountering technical topics that were new to me like advanced composites, semiconductors, and digital signal processing. These were areas I had not had the opportunity to learn previously. While an aeronautical engineering background provides good depth, aerospace systems require many disciplines to come together like electrical, civil, mechanical, and software engineering. This degree allowed me to open my aperture and learn topics that helped me be more effective in designing space systems.

My third advanced degree is a Technology Management MBA from the University of Washington. About eight years into my career, I began to realize that I needed to understand a new language, that of the business world. As I advanced in my career,

I went from doing technical engineering work to also trying to get my projects funded by writing proposals to customers for contracts. Trying to get that funding came with internal company inquires to answer like net present value, nonrecurring engineering cost, and return on investment. At the time, I had no idea how to calculate those values or even what the inputs to the equations were. I decided that if I wanted to get my projects funded, I needed to learn. I entered into an intensive 18-month MBA program designed for working technical professionals. I learned the language of business and even more about frameworks for understanding leadership, economics, and marketing. With my MBA, I never intended on changing careers into finance or sales. I wanted to understand business so I could be a better engineer and leader.

Powering On

Was there a significant event that changed your career trajectory and what was it?

There have been a couple of times in my career I have taken a bit of a leap of faith. Those leaps were opportunities presented to me by people with whom I developed relationships with at work over the years. Both conversations ended with the person presenting the opportunity saying something like, "I cannot tell you exactly what you'll be working on, but I think you will really like it." Because of the proprietary nature of some aerospace projects, not knowing exactly what you'll be working on before accepting a position is not uncommon. Because I knew both of these people, I was willing to take a leap.

The first opportunity came after nearly three years of working in Boeing Commercial Airplanes (BCA). I had gotten involved with an informal group of engineers working on vehicle configuration design throughout the company. We would get together once a month off-hours and share presentations about the projects we were working. As I learned more about the defense and space side of Boeing's business, I found it was interesting to me and was more in line with the work I had wanted to do when I first became interested in aerospace. Eventually, the leader of that group asked me if I would like to go work with him. He became one of my mentors for many years. I transitioned to Boeing Defense, Space, and Security where I spent the next fifteen years working on preliminary vehicle design and rapid development of advanced aircraft and spacecraft. It was fascinating and rewarding work.

The next opportunity also presented itself through someone I knew. He worked at Boeing Defense, Space, and Security previously, and I knew him from the time I was moving from BCA to that group. Many years later, he went to work for Blue Origin. I had been aware of Blue Origin, but because the organization was not publicly communicating about their projects at the time, I did not have details. I had applied and initially declined an offer from the company. He asked me to reconsider the company's offer and to come work with him. I eventually did decide to accept the offer, and I became one of the first people hired to work on the New Glenn program full time.

The relationships I have developed with people throughout my career have been vitally important. Those relationships can be very long lived, coming back to the forefront years later. The people in my "leaps of faith" gave me opportunities, and I have also tried to do that for others.

Ginger Shao

Senior Director (Retired)
Honeywell

About the Author

Ginger Shao joined Honeywell, a Fortune 100 company in the Aerospace industry in 1997, from an academic position. She was in engineering and management leadership positions and retired from the company in 2020 as a Distinguished Fellow, the highest technical ranking in the company, and a Senior Director. Her responsibilities included strategy definition and product roadmap development on data analytics, advanced technology development and solution design for analytics new product introduction, investment evaluation for analytics technologies, and analytics technology portfolio management.

She specializes in proposal development for new pursuits from both government and industry opportunities, systems engineering of aerospace safety-critical systems, modeling and analytics of physical systems with digital twins, machine prognostic health management, and technology investment and management. The projects and opportunities she worked on ranged from millions to hundreds of millions in dollars. She has multiple publications and was awarded multiple patents from industrial applications. She has a PhD in Simulation and Modeling of Atmospheric Systems from Colorado State University and an MBA from the University of Arizona and is Six Sigma Black Belt certified.

She served as the founding Chair of the Digital and Data Steering Group, which was in charge of creating standards committees to develop digital and data-related industry standards. Included in those committees were the G-31 Committee (Electronic Transactions in Aerospace) and the G-34 Committee (Artificial Intelligence for Safety-Critical Systems). She also served as Vice Chair of the IVHM Steering

Group as was a Session Chair multiple SAE conference meetings. She was the Co-Chair of the PHM Society Annual Conference in 2019.

She is currently serving as Advisory Board Member of the Department of Aerospace and Mechanical Engineering at the University of Arizona and Board Member of an investment company. She also owns a consulting company for professional services and asset analytics for industrial applications.

In her spare time, she enjoys dancing, sports, reading, music, and traveling with family and friends. The picture below shows her (center) traveling with her dancing friends on Lake Powell in Arizona.

Ginger Shao

Questions and Answers

Personal Career Insight

How did you decide between a leadership vs. technical career track?

Deciding on a career track is what most of us would face in our early careers. It's the result of the gradual understanding of ourselves as well as the industry in the early stages of our careers, an inevitable step in our growth process.

The decision for me to start with a technical career was straightforward—I had a PhD degree in Science, and being an engineer when joining the Aerospace industry was a natural choice for me. I was always curious about things and liked to explore unknowns, so having an engineering problem in hand to solve kept me motivated all the time. Because of that, I really enjoyed the decades of my career in the industry with a Fortune 100 company, Honeywell. I joined the company as a junior engineer and retired as a Distinguished Fellow, the highest technical ranking in the company.

However, I did struggle in the early years as for how I would choose my career in the long run, and actually I had quite some detours." I started to look for more challenges after a couple of years with the company and got exposure to the Aerospace AT management, dealing with a multi-million dollar technology investment portfolio. I gradually worked into the portfolio manager position, leading the process for the technology strategic planning, product roadmap alignment, waterline definition for R&D project investment, and IP protection. During the transition process, I earned my MBA degree and was certified as a Six Sigma Black Belt.

While the management track looked so promising, a family constraint limited my potential in advancing to senior leadership positions. After some careful considerations, I decided to return to the technical role as an individual contributor. Transitioning back was smooth for me due to my technical background by training.

There is one key lesson that I learned through my experience: don't take your career track as a linear path, but adjust as you grow or as opportunities present themselves. When we were at our earlier career stage, our visibility might be limited by what was offered to us, like the career tracks with the steps on the ladder defined by the company. While you were in engineering, you might not be looking into business management, manufacturing, or supply chain. As I joined the company, I might have thought that a track to Technical Fellow was where I was heading, but I didn't imagine that I would be led into a management position through a Six Sigma project and would have the courage to tackle an MBA degree while working full-time. I also didn't realize how critical mobility was in pursuing a senior leadership role, which ultimately led me to decide that being an individual contributor was a better choice for me for a work-life balance. But they all happened and gradually evolved as I learned more about myself and was driven by the situations.

Work/Life Balance

Did you ever have to make a move to advance your career (within your company or changing companies) that impacted your family life, and how did you balance the two?

Not being able to move has been a constraint for my whole career, which dictated my career path and determined my directions at the two critical turning points—one was from academia to industry and another was from management to individual contributor.

I naturally started with an academic career when I got my PhD degree, posing for a professor position. As a couple years passed by, the constraints on my movability became obvious: my husband was in a professor position and he was not very movable. I had to turn down offers that required me to move. The tough question for us was; given our first child was only a few years old, should I move and leave them behind, or take the child and leave him behind, or let him quit and follow me? We couldn't get the question answered, and after several opportunities, I decided to quit pursuing an academic career but to turn to industry, which would have a better chance of not requiring me to move. I was quickly hired by Honeywell and worked there until I retired.

When I was with the company, I grew into a management track, to the point again that I had to move to take a senior leadership position. We ran into the wall again. I did an earnest assessment of my situation and decided to come back to the technical track, which I kept advancing from there.

As I look back, I am happy with my decisions. Besides my personal achievements, our two children have graduated from Stanford University and have become young professionals. There is nothing more rewarding than having a happy family with achieved children, Other alternative choices could have turned out fine, but I feel blessed to be where I am.

Movability is no doubt one of the biggest challenges for career advancements, especially for women. What we need to do is honestly assess our priorities, strengths and constraints, and find a position that fits best into our situation. It is by no means easy, and oftentimes the decision is very personal. Recent technology advancements allow more opportunities to work remotely, but as far as I can see, some key positions may still need co-locations. When that happens, we may need to take one step at a time and explore options allow the way, especially when our family is part of the equation.

Mentorship/Sponsorship

How important was mentorship/sponsorship for your career? Have they been men or women? How was the relationship established?

I greatly benefited from mentorship, and strongly recommend it for people at any stage of their career. It enhances personal growth, increases work efficiency, and adds to job satisfaction. I experienced three stages of mentorship along the maturation of my career: expansion on knowledge and business operations, targeted leadership, and leading for impact.

The first stage was when I just joined the aerospace industry from academia. Learning how a company works and how to effectively convert my PhD-level technical skills to the company's product applications were the first challenges. I had very limited ideas about the functions within the company and how the company operated. I even didn't know the ranking of my position on the career ladder but simply took what I was given when hired. Team members often guessed my position two or three levels higher than what I really had due to the challenging technical assignments I was took on. In this phase, I had mentors from the manufacturing floor, product subject matter experts, and management. I asked them to be my mentors per the functional needs I felt I needed for knowledge and guidance. They advised me based on my capability, potential, and needs. I grew quickly and was promoted every year in the first three years of employment, partly due to my hard work, and inevitably due to the product and operational knowledge that I could quickly gain from my mentors in different areas.

The second stage was when I was identified as a high-potential employee and was selected to be on the fast-track to a leadership position. I was positioned for technology management and got exposure working with C-suite leadership on critical technical issues, including technology investment and IP protection. The work scope required me to work closely with senior leadership teams from business, legal, and contracts, besides technology and product development. My mentors came from both business and legal backgrounds. This stage was a boost to the rest of my career in the company when I could see things from multiple angles, and it made me a much more efficient decision-maker. Again, I proactively pursued mentorship for the areas I needed to grow.

The third stage was when I was in the leadership role as a Technology Fellow. Before I took the Fellow position, I shadowed another seasoned Fellow for four years, learning the key practices and responsibilities. I was introduced to the key players and processes, which enabled me to transition smoothly into the position when the time came. I also had senior leaders as my mentors in guiding me on the strategic directions the company was driving towards so that I could effectively lead technical efforts for alignment. My mentors also served as great sounding boards of my ideas so that my work was more impactful to outcomes. At this stage, I got my mentors with recommendations from management, who knew much better the strengths of individual leaders and made good matches to my needs.

My mentors were both men and women. Mentorship does not necessarily need to be from a formally defined mentorship program, although it would definitely help. Mentors are the strong supporters along the way of our continuous improvements. From that perspective, mentoring is really a need driven from within. We would need to have an honest assessment of ourselves and then reach out every which way to find mentors to help us grow. I have been on both sides of mentorship, and I can testify that it has been a joy in bonding with people to learn and to share.

Avoiding a Stall

How important is an advanced degree?

Having a bachelor's degree is likely a starting point for high-tech companies these days, including Aerospace, so I assume that's not a question for debate here.

As far as the value of pursuing a master's or a higher degree, the question really comes down on how you'd value the classroom learning versus learning through working experiences, and how it may impact career selections and growth.

If one can afford the time and money, I would suggest getting an advanced degree as much as you can. An advanced degree allows you to gain more specialized knowledge and critical thinking skills with rigorous training, which enables qualifications for opportunities with higher entry barriers, as well as broadening the depth and breadth of the scope that you may take on with your work. With research experience, it may also allow you to have more flexibility in choosing different roles in the organization. As an engineer, one can work on product research and development with front-end technologies, or product development with mature technology applications.

My career definitely benefited from both of my PhD and MBA degrees. The PhD training provided me with solid math and physics background, and independent problem-solving skillsets. The knowledge and the skillsets made me ready to take on the product analysis and modeling tasks from the first day of the job and allowed me to independently make an educated assessment of project challenges and to provide proper solutions. The knowledge and associated systematic thinking also made it easier for me to understand various product design insights, which greatly sped up my learning of product domain knowledge. The critical-thinking skills benefited me in every aspect of my work. The MBA degree, on the other hand, opened another dimension in my "polarized" technical brain, bringing in a brand new way of thinking and perspectives in working with people and business. I considered it fortunate for me to have both rigorous trainings at the early stage of my career, which provided me with great flexibility in career choices and advancements.

With that said, we definitely need to recognize that learning is a life-long process, and learning through real-life working experiences is the majority part of our growth journey. An advanced education can train us with strong self-learning skills, which provides a solid foundation for our continuous growth; however, becoming an expert and a leader in a specific discipline requires continuous learning and determination.

Powering On

Did you ever feel limited in a role and how did you circumvent that?

I joined the industry with excellent academic skills but shortage of knowledge in industrial products and business processes. I started with high-challenging technical works, while catching up on the industrial practices during the process. After a couple of years immersing myself in the industry, I was ready to move on to pursue new opportunities. I stared with the strength and gap assessment between where I was now and where I saw I could be.

Due to my research experience, I looked into R&D and technical management roles. As identified as a high potential employee, I took the opportunity from the company-sponsored MBA program, while working full-time. Those two years were an endurance test on both my physical and mental strengths. Besides the full-time job and MBA curriculum, I also had my second child, who was only one-years old. As I look back, it is amazing how much a person can accomplish when keeping the light at the end of the tunnel in mind. Apparently the hard work created mutiple potential paths that I could move forward with my career. As stated earlier, I went into management, an experience that benefited my career for the decades to come.

As we learn more and more from work and about ourselves, it's inevitable that we often grow out of what we have been doing. What I learned is that we need to take the responsibility for ourselves to shed off the old skin and grow into our newer and bigger skin. We need to be honest with our strengths, limitations, and motivations. We need to know what we want, and how we may get there. Some take hard work, others take proactive pursuits. Having a new degree, as in my case, is n extreme example, while mentoring and networking could provide other alternatives in generating opportunities. Make your wishes know so that people can help.

Bridget Sheriff

Vice President of Engineering
for the Interiors Business Unit
Collins Aerospace

About the Author

Bridget Sheriff is vice president of engineering for the Interiors Business Unit of Collins Aerospace, a division of Raytheon Technologies, based in Winston-Salem, North Carolina. In this role, Bridget is responsible for all engineering and technology activities to support the business' portfolios: Seating, Interiors Products and Service and Support. This is a group of over 2500 engineers at 20 global locations who design and develop Interior products and aftermarket services. Bridget enjoys creating global teams that connect people across organizational boundaries to solve difficult problems. She sets high expectations for her teams to share and propagate best practices, encourages talent development, and leads a culture of innovation built on respect for diversity.

Bridget began her career as an electrical design engineer at Honeywell. Then she served in a variety of engineering roles with Hamilton Sundstrand and UTC Aerospace Systems, including software engineer, engineering manager, senior director of product line engineering, and executive director of electronics and embedded controls. Through these roles, she has developed her leadership skills as well as a passion for applying model-based systems engineering methods not only to product solutions but also to development lifecycle logistics.

Bridget serves as Raytheon Technologies' executive sponsor of North Carolina A&T and actively promotes Collins wide engagement with various STEM organizations like Girls Who Code, Introduce a Girl to Engineering, and Invention Convention. She also sits on the Collins Corporate Social Responsibility board.

Bridget earned a Bachelor's degree in Electrical Engineering from the University of Illinois Urbana-Champaign and an MBA from the University of Wisconsin—Madison.

Questions and Answers

Personal Career Insight

How did you decide between a leadership vs. technical career track?

Early in my career (like 6 months), I was given the responsibility to lead a product development. I hadn't given career planning any consideration at all, but I quickly learned that leading with no technical proficiency was not a recipe for success. Although I learned a lot, I was not satisfied with my performance in the role and had no ability to be pragmatic. In my next role, I was an individual technical contributor. I focused on building a foundation of experience that I could reach back into and establishing technical credibility—not really for my reputation, but for myself.

I held technical individual contributor roles as well as project leader roles. I really enjoyed being part of a team and getting involved in solving difficult problems. I learned that, rather than work alone, I am better at extracting ideas from others and organizing or coordinating those with the rest of the team to get to answers. Most importantly, I learned that my personal motivation is strongly associated with my team's engagement and support. I would do anything to help my team's performance.

I could see that there were others around me who had higher technical capability and I migrated to the leadership track. As I moved into more leadership roles, I was very interested in figuring out how to extract individual and team potential. As I took on more leadership responsibility, I learned that my reach was extended to connect more people and influence bigger outcomes. I love seeing people happy about their accomplishments, engaged in their work, and benefiting the business. I still get to be involved in plenty of technical challenges, but I know I've chosen the right path for me.

Work/Life Balance

Did you ever have to make a move to advance your career (within your company or changing companies) that impacted your family life, and how did you balance the two?

Yes, I did. Since graduating from the University of Illinois, I had lived and grown my family in a small town in northwest Illinois. My career progressed very well over 17 years at a single location for United Technologies. I still had runway in that location, but my boss and mentor, Bill, saw a different path for me to grow and "leave the nest." This meant relocating to Arizona, far away from relatives and friends. It also meant that my husband's career came in as second priority. Honestly, it wasn't very balanced—the move benefited me more than it benefited anyone else in the family. All relocations are very stressful and I couldn't and wouldn't have made the move without the support and buy-in of my husband. We talked through the opportunity and he stood behind me. He has always been one of my biggest fans and has encouraged me to keep moving up in my career.

Bill's vision for me was spot-on! My development absolutely accelerated when I left the nest. After the first cross-country move, the next one was much easier and again propelled my growth and career trajectory.

Mentorship/Sponsorship

If you had a Sponsor to reach executive levels, how did you meet that person? Was the person assigned to you or was it a boss that became your sponsor? How did this person help your trajectory?

I knew that I was labelled a "hi-pot" and that there were "people above me" who were advocating for my career—but I didn't always know who they were. However, Bill, my long-time sponsor, was certainly active in helping me to get into many roles, including my first executive role. I met him when I was working on a very difficult assignment in my first real leadership role. Bill always tells the story that since he was sent from headquarters to "help," I didn't trust him when we first met. He quickly gained my trust and became my coach and mentor, not as an assignment, but because he saw my performance and my potential. He helped my career trajectory by constantly challenging me, having confidence in me, and never being afraid to deliver tough feedback. Bill was my boss for a while, but even when he wasn't, he advocated for my career and drove me into opportunities to grow beyond my comfort zone. I have often felt that Bill believed in my career prospects more than I did. His coaching still helps me tremendously.

Avoiding a Stall

How important is an advanced degree?

It might not be a popular answer, but I would say that getting an advanced degree for a leadership role is less significant than continuous learning—which is vital. Keeping yourself up-to-date on relevant topics, recognizing your gaps in knowledge or skill, and then taking the initiative to fill those gaps—is a differentiator. I went back to school 20 years into my career and got an MBA. I could see that I was missing some business acumen that would benefit my current job performance and would be required to excel in any future growth roles. I didn't go back to get the advanced degree—I went through the program to learn the skills and shore up my knowledge gaps. Of course the credentials never hurt!

Powering On

Did you ever feel limited in a role and how did you circumvent that?

One of my roles was to lead a centralized engineering organization. Most of my time was spent on very tactical activities. I felt limited in the role because I couldn't see where it was going. The vision, purpose, and future for the team was not clear.

I circumvented that by assembling a team to create it instead of waiting for it to be passed down. We were empowered and enlisted the buy-in of key stakeholders to create the vision and purpose for the organization. After we created an initial version and a preliminary roadmap, I no longer felt limited. I knew how the tactical activities aligned to a higher purpose, and I felt empowered to lead strategically towards the vision.

Amanda Simpson

Vice President, Research and Technology
Airbus Americas

About the Author

Amanda Simpson was named Vice President for Research and Technology at Airbus Americas in 2018, responsible for coordinating technology development, research activities, and innovation for Airbus in the western hemisphere.

Ms. Simpson joined Airbus following government assignments in the United States (U.S.) Department of Defense. She was the Deputy Assistant Secretary of Defense for Operational Energy, responsible for developing the strategy for the utilization of energy for military operational forces worldwide, and was the senior advisor to the Secretary of Defense for all matters pertaining to energy in our military. Prior to accepting that responsibility, she was the Executive Director of the U.S. Army Office of Energy Initiatives, an organization developing large-scale renewable energy projects to bring energy security to Army installations leveraging private sector financing.

Previously, she was the Special Assistant to the Army Acquisition Executive. In that role, she was a principal advisor to the Assistant Secretary of the Army (Acquisition, Logistics, and Technology) on all matters relating to Army acquisition, procurement, research and development, and logistics. In 2010, Ms. Simpson was appointed by President Barack Obama to the position of Senior Technical Adviser to the U.S. Department of Commerce where she advised on policy and export control issues necessary to protect the security of the United States, making her the first openly transgender female presidential appointee in U.S. history.

Prior to her government appointments, Ms. Simpson was the Deputy Director for Force Protection in the Advanced Missiles and Unmanned Systems Product Line at Raytheon Company Missile Systems in Tucson, Arizona. She held several program management positions on technology development activities for missile systems sensors and development programs.

Previous to her tenure in program management, Ms. Simpson was the manager of captive flight test operations for Raytheon Company and Hughes Aircraft Company, which included the duties of mission director and project pilot on numerous developmental test programs. Her team received the DARPA Award for Significant Technical Achievement in 1999.

Ms. Simpson has been the recipient of numerous awards and recognitions including the 2004 Tucson YWCA Woman on the Move, the 2005 Arizona Human Rights Fund Individual Award, and the 2015 National Conference for College Women Student Leaders Women of Distinction Award and was named an Outstanding Alumni of Harvey Mudd College in 2018. She is a recipient of the Secretary of Defense Medal for Outstanding Public Service and the Department of Defense Pride Civilian Leadership Award. She is a nationally renowned speaker and has presented at corporations, government agencies, civic organizations, conferences, and colleges around the country on gender and diversity, as well as technology and aerospace innovation.

Ms. Simpson earned a Bachelor of Science in Physics from Harvey Mudd College, a Master of Science in Engineering from California State University, and a Master in Business Administration from the University of Arizona. She is an Associate Fellow of the American Institute of Aeronautics and Astronautics, serving on the Systems Engineering Technical Committee, Executive Steering Committee, and Corporate Member Committee. She is a Fellow of the Royal Aeronautical Society and serves on the

Amanda Simpson

board of the Washington DC Chapter. She serves on the board of directors of the Commonwealth Center for Advanced Manufacturing and Advanced Thermal Batteries Inc, and is the chair of the board of directors of the Airbus Institute for Engineering Research.

Ms. Simpson holds both an Airline Transport Pilot certificate and a Certified Flight Instructor license and has logged nearly 3,000 hours of flying in more than 60 different types of aircraft including floatplanes, flying boats, unmanned drones, helicopters, light aircraft, and multi-engine jets.

Questions and Answers

Personal Career Insight

What was the coolest thing you experienced in your career?

Being a huge fan of aviation, I've had some wonderful experiences flying aircraft that I could never have imagined when I began my career. Let me share a few:

Following a discussion at a conference with someone who knew someone who knew someone, I received a call from the manager of captive flight operations inquiring if I would be interested in joining his organization. He informed me that besides an interview to determine if I would be a good fit, the application process included a flight in the company jet to determine if I had "the right stuff." A week later, after reading the flight manual the night before, I was off in a Sabreliner business jet with a couple of other pilots on a recurrency flight. A half-hour later I was invited to the cockpit and asked to sit in the copilot's seat. In the left seat was the company's chief pilot. He had me fly straight and level, do some level turns, and then asked for a recovery from an aerodynamic stall. Up to this point, I had only flown single piston engine aircraft, and the procedure was to recover to the best climb speed, which I did in the Sabreliner. However, with two jet engines, this pointed the nose high in the sky and the climb rate was several thousand feet a minute. The chief pilot chuckled that I was heading for outer space. I got the job.

Years later, in that same aircraft, which was modified for a test project, I believe I set an (unofficial) world record for piloting a jet aircraft below sea level. Collecting data for the Navy over the calmest water I could find, we flew 50 feet above the Salton Sea in the southern California desert at dawn (before the wind came up). We did this for nearly two hours per flight, twice a day for several weeks. I don't know that anyone could top that.

I had the opportunity to fly in close formation with active military aircraft (and when I say close, it was within a few feet), chase un-piloted drones in military ranges, and get some astonished looks from other pilots or airplane watchers when I popped a drag chute upon landing or folded the wings on an A-3 while taxiing. I drew astonished looks from military crews when a tall blond stepped out of the aircraft or attended pre-mission briefings at bases. All enjoyable memories.

But one career event stands out above others, and it didn't involve flying. Later in my government career, I was invited to attend a reception at the White House and, upon entering the main room, was informed they had been looking for me as the President of the United States was waiting for me. That was cool.

Work/Life Balance

Did you ever have to make a move to advance your career (within your company or changing companies) that impacted your family life, and how did you balance the two?

In the summer of 2009, I was struggling with a decision that would indeed impact my career path as well as all my personal and family relationships.

A few months earlier I had been contacted by the White House to determine my interest in working for the Administration. Unfortunately, that opportunity didn't work out as the hiring manager wasn't comfortable with my lack of experience inside the Pentagon. Now another opportunity was offered, but it was outside of my expertise and thus outside my comfort zone, but was advised future opportunities could become available if I entered into government service at this time. As I had a planned business trip to Washington DC on my calendar, I arranged to meet with the nominated, but not yet Senate confirmed, Under Secretary. He very much encouraged me to join his team.

At home, my teenage son, whom I shared custody with my ex-wife, had recently started high school. While during middle school he was spending half his days and nights at my house, he felt it would be easier to base out of one home at this age, and my home wasn't serviced by the high school bus. However, I believe he was getting to that age when he felt his friends wouldn't understand that he had two moms (now divorced) and that one was his father. We had several conversations that summer and he encouraged me to pursue the position in Washington.

Other than my son, my family resided just a short hour flight from Tucson in California so DC would be a longer flight to visit, phone calls were still the primary mode of communication amongst us.

But in addition to my family, it was friends and the community that had supported me that was part of my evaluation. I had lived in the Tucson area for over 15 years, made friends with coworkers for over 20 years, and the GLBT community center had been a huge support when I transitioned my gender a decade earlier. I was active in community organizations, my homeowners' association, and had strong connections with the local aviation community. There were even friends from college that lived in the area. The impact on all these relationships weighed heavily upon me.

In the end, I had to determine what would be best for all, setting my own personal issues aside, I took the position in the Administration. Return visits to Tucson to visit family and friends occurred several times a year, and my son spent a few summers in DC, even interning for a Member of Congress for several months. Moving alone to a new city where I knew nearly no one was challenging, but I understood that my participation in our government would be impactful and that personal issues would find a new balance over time.

Mentorship/Sponsorship

What activities have you engaged in that have helped other women achieve success in their aerospace careers?

Long before I transitioned, I was assisting others to achieve success in their careers. Whether it was formal mentoring programs or informal discussions with younger colleagues regarding career development, many knew that my office door was open. (That's a figurative office door as the company had put everyone in cubes upon consolidation in the early 90s.) I didn't make obvious preferences between men and women as I was scared that someone might deduce my secret, but I seemed to have a higher proportion of women mentees than the general workforce.

Following my gender transition, I became quite active in the women's employee resource group and served on the board for a few years.

When asked, I would share my observations from the "other side" to assist in understanding the career environment and offering insights into how to be successful. I would present concepts I learned in management classes or explore ideas from books in workshops at the annual women's conference. I not only mentored more women, but reverse-mentored male members of the company's leadership. When considering candidates for open positions, I always ensured that those selected for interviews were balanced and encouraged women who personally may not believe they were qualified or ready for assignments to pursue opportunities that would advance their careers.

At forums and summits sponsored by professional organizations where I held membership, I would routinely offer to participate in diversity panels and mentoring sessions for aspiring youth, and work with organizing committees to ensure there was female representation amongst panel and keynote speakers.

I've been in positions where I am seen as a role model, but it's equally important to continually celebrate the successes of others, promote them to other leaders, and follow up with mentoring and career advice. I continue to pursue many of these initiatives today at my current employer and use my executive position to influence other departments to institute similar progress.

Avoiding a Stall

With respect to your career, did you ever hit an organizational roadblock? How did you overcome it?

A couple of years after my gender transition, I had a meeting with the Director overseeing the Special Projects. At that time I was managing several small projects, each running a few months to a year, and when each was concluded to the customer's satisfaction, I was assigned a new project to manage with an appropriate team to execute. However, in the previous months, I had successfully completed a few projects but did not receive new assignments to take their place. Not only was I not busy at work, I was getting bored.

I shortened my days back to a 40-hour workweek, was completely inefficient with my work time, and decided I would run for public office in my now excessive spare time. This of course led to opportunities years in the future that I couldn't imagine at the time, but I felt underutilized at work. Not only was I bored, I was frustrated. So I met with the Director.

He told me "There are people uncomfortable with what you did, and they don't want to work for you. I can't make them work for you, so I can't give you any more work." I must admit that I walked out of his office a bit dazed and confused as it had been years since my public gender transition and thought everyone had moved on. Many had but, apparently, not all.

Thus I went searching for work inside the company. I talked to a few other Directors and eventually found a position as a chief engineer and lead architect on a large development activity in Connecticut. While this required me to be away from home for two weeks at a time, it did impart unique experience and expose me to connections that were useful on the next steps in my career path. Later, to get back into project management, I took on anything that was available, and that primarily was high-risk, low-probability-of-success assignments. But because of my experience in building high-performing teams and my ability to understand the new technologies, I was highly successful.

I eventually discovered who those "people" were and was pleased when some of them asked to work on my teams, as by then I was perceived as the manager to work for if you were interested in cutting-edge technology and exciting opportunities. It was no surprise to me that the Director who wouldn't offer me assignments was told to accept an early retirement as his management practices were not aligned with company policy.

Powering On

Was there a significant event that changed your career trajectory and what was it?

With little doubt, the decision to move forward with my gender transition changed my career and life trajectory.

I vividly remember making the appointment with the president of our multi-billion dollar defense segment to tell her I was going to transition my gender. Showing up for that meeting was indeed an irreversible milestone on my journey. It was one of the most difficult conversations I ever had to initiate. I had a hard time just talking. I was afraid that I would lose everything. Fortunately, we had known each other well for over a dozen years and helped me get through the discussion.

She did offer to relocate me anywhere within the parent corporation with the same pay and responsibility. While it was a nice offer as this was the standard option offered by companies, I wanted to stay at that facility as I really enjoyed what I did as well as the people I worked with. In addition, my young son and my support structure were in that town so my hope was to be able to transition in place. We joked that I would have to take a 27% pay cut as that was the average difference between the genders. But she did empower me to work with our human resources department to put together a transition plan for those I worked and interfaced with. I knew what I needed to do as I'd been thinking it through for months, but for many it was going to be something they had never contemplated, and there needed to be a plan in place.

Sure others had transitioned before at the company, but there were no policies, no best practices in place. There had never been anything written down and few even in human resources knew that others had transitioned before. I knew from my years of managing teams that documenting the process and policies would be important so that others in the future may benefit from what I and the organization would learn and experience. And it was that plan, which I developed with a close friend, that was adopted into practices that were eventually shared and embraced by most corporations over the following decade and is still used as a guideline today.

I firmly believe the saying "the only person that I can change is myself, and sometimes that makes all the difference." The steps initiated following that meeting had me change my focus from flight test to program management, which would later lead to participation in local and state government, and eventually a position in a President's Administration. Some were planned, others were inspired by opportunities, and a few were the results of making compromises to compensate for the lack of understanding and empathy in others. But life is a culmination of the reactions we make to the world around us. I am quite fortunate that the result has been positive.

Karen Taminger

Senior Materials Research Engineer
National Aeronautics and Space Administration
Langley Research Center

About the Author

Karen Taminger is a Senior Materials Research Engineer at the National Aeronautics and Space Administration (NASA) Langley Research Center in the Advanced Materials and Processing Branch. She currently serves as a technical lead for structural efficiency in one of NASA's transport aircraft projects, and she has led additive manufacturing work for several in-space and for-space manufacturing projects. Throughout her career at NASA, she has worked on various aspects of processing-microstructure-mechanical property characterization related to metallic materials for aerospace structures. She has led the development of the Electron Beam Freeform Fabrication (EBF[3]) technology since 1999, a large-scale metal additive manufacturing process for high-performance, low-cost fabrication of metallic structures for aircraft, launch vehicles, and spacecraft.

One of the highlights of her career was the opportunity to conduct parabolic flight testing of EBF[3] for compatibility with the space environment. As a result, she has spent 3 hours in zero gravity (in 15-second increments!). She is the co-inventor on five issued patents and four other patent disclosures and was the lead or co-author on more than 35 papers and 100 presentations. Personal awards include the NASA Exceptional Achievement Medal in 2007 and the NASA Exceptional Technology Achievement Medal in 2014. Her team's work in metal additive manufacturing was recognized with the NASA Langley's Whitcomb-Holloway Technology Transfer Award in 2008 and selected as the runner-up for NASA Patent of the Year in 2016. She earned her Bachelor of Science with Honors from Virginia Tech in 1989, majoring in Materials Engineering, and her MS degree from Virginia Tech in 1999 in Materials

Science and Engineering. She was employed at NASA Langley Research Center as a cooperative education student from 1986 to 1989, and she has been there full time since graduating with her BS in 1989. She is married and has three sons.

Karen Taminger

Questions and Answers

Personal Career Insight

What inspired you to choose a career in Aerospace and what has given you the most satisfaction during your career?

I enjoyed mathematics and chemistry in high school and was intrigued by the bourgeoning world of computers (it was the early 1980s). However, I was skeptical when my guidance counselor recommended I study engineering. You see, when I was growing up, my father was in the Navy so my introduction to engineering was in the boiler rooms on his ships. It was a hot, challenging environment, and women were not allowed to serve on ships at that time. My mind was changed the summer between my junior and senior year in high school when I was selected for an eight-week program at NASA Langley Research Center. I was assigned to do computer programming of algorithms to analyze wind tunnel data, but I was also immersed in a world of aerospace engineering. Throughout that summer, I spent my spare time visiting the worksites for the other students in my program. I learned I wanted to do more experimental work than computational or theoretical, and that there were many different types of engineering to choose from (none of which required working in a boiler room!).

I decided to major in materials engineering and was hired as a cooperative education student at NASA Langley Research Center early in my undergraduate studies. After graduating with my Bachelor of Science degree, I was converted to a full-time engineer and have been at NASA ever since. I have had the opportunity to work on many different projects throughout my career. One of the most satisfying ones involved developing a portable metal 3D printer for performing tests in zero gravity. The project focused on demonstrating the feasibility of electron beam/wire directed energy deposition in a zero-gravity environment. We conducted these experiments in NASA's C-9 aircraft, which flew in parabolic trajectories to create periodic simulated microgravity conditions. Besides the thrill of experiencing zero gravity, the greatest satisfaction came from building this entire project from advocacy to demonstration. I had to start with obtaining the funding and assemble a team to design and build the hardware. The camaraderie, focus, and dedication of the team was inspiring, and we overcame numerous hurdles to successfully perform a series of cutting-edge experiments on metal additive manufacturing in microgravity more than a decade ahead of everyone else.

Work/Life Balance

Did you ever have to make a move to advance your career (within your company or changing companies) that impacted your family life, and how did you balance the two?

I have always taken a slightly different path from my peers. For me, changes to my family life came first, and that move enabled me to advance my career. My husband and I are both engineers, and after the birth of our second child in two years, we were exhausted from trying to maintain two careers and two kids. We did the calculations and realized that one of the two of us was basically working to pay for the daycare bills and other services like maintaining our cars, mowing the lawn, and take-out meals—all things that we could do on our own but didn't have the time. Most importantly, we were too tired to enjoy raising our children. Since my job had better benefits, my husband made the sacrifice to put his career on hold and become a stay-at-home-dad. The decision was not made to advance my career so much as to put family first, but that has made all the difference.

The outcome of that decision took me to places I never envisioned at the time. With the children at home (we added a third three years later) and my husband there to nurture them, I could take on more responsibilities without having to worry about getting the call to pick up a sick child being sent home from daycare or school. I began to travel more and to develop a wider technical network inside and outside of NASA. I tried hard to balance meetings and trips and not to miss events like sports games or school assemblies, which sometimes led to crazy travel schedules. I was comforted knowing that my husband was there with my three sons, videotaping the occasional event that I did have to miss. That support at home gave me the stability and confidence to take on the challenges of leadership at work.

Mentorship/Sponsorship

How important was mentorship/sponsorship for your career? Have they been men or women? How was the relationship established?

Mentorship, both being mentored as a young engineer and now providing that mentorship to young engineers in our group, has been key to learning on the job and navigating how things get done within a big, ever-changing organization. As a co-operative education student at NASA, I had formal mentors assigned to me. In the co-op program, I alternated quarters between taking classes at Virginia Tech and working at NASA Langley. I spent five quarters at NASA before graduating with my undergraduate degree. Since I was a student, I was learning the practical aspects of how we performed different tests on different materials. I rotated through different branches covering the breadth of disciplines within materials engineering. For these mentorships, three of my five different mentors were female engineers.

After graduating with my degree in Materials Science and returning to work at NASA full time, I found the nature of my mentorships and my mentors changed. Although we have many formal procedures, policies, and training on how to follow them, there are many aspects of engineering that are not written down. Much of what I needed to know to do my job effectively was informal and so were the mentorships that helped teach me along the way. I learned about programmatic advocacy, how to manage people, and team dynamics from my branch management and technical leaders, all men 20-30 years senior to me. They also helped me learn how the "system" worked and develop connections that have enabled me to grow my network and influence. In addition, I was fortunate to have several female engineers as colleagues only 5-10 years older than I am. Beyond technical mentoring, I learned a lot about personal growth and balancing family and work from them. They also provided me with an outlet for comparing notes about our children's development. I have now progressed to being a mentor to our new hires, both young men and women, to share with them what I have learned along the way.

Avoiding a Stall

Were you ever presented an "opportunity" you declined and did it hurt your career? If so, how did you overcome any negative impact?

NASA has a dual ladder for advancing to more senior positions: this means you can attain promotions by continuing on a technical track or by switching to either a program or line management position. About fifteen years ago, I was a mid-career engineer, leading and performing research on metals processing on the technical side of the ladder. I had worked with project managers for several years providing technical support, so I had learned how the project offices at NASA work. We had an opening as an assistant branch head in our line management track, so I applied. After discussions with the branch head about what the duties would entail, I had to make the choice between continuing to support the projects and personal research that I was performing or switching over to line management where I would be more involved with personnel development. I had already attained the same pay grade in the technical ranks as I would have received as an assistant branch head, but branch management is a stepping stone to higher levels of either program or line management. I opted to withdraw from the applicant pool to continue the work I had spent years cultivating.

Although I've preferred to continue to lead from within and have immediate technical influence over the programs and research, I find that I have now been doing more personnel development in that role. I have always mentored summer interns and am now mentoring new employees. I have been able to help build the pipeline and groom young talent as many of our branch members are reaching retirement age. In my technical lead position, I can advocate for funding and new positions that will form the future workforce within our organization without actually being in branch management. I may not be eligible for a senior manager position without becoming a program or line manager, but I firmly believe that there is also the need for the strong leadership from within that I provide. From that perspective, I do not feel as though there has been a negative impact on my career by opting not to pursue the management track.

Powering On

Was there a significant event that changed your career trajectory and what was it?

I started my career at NASA as a materials research engineer performing mechanical tests, microstructural analyses, and evaluating environmental effects on various aerospace metallic materials. In 1999, a new program was started at NASA Langley offering support for exploring nascent ideas with promise for advancing technologies relevant to future aircraft and space-based applications. At that time, the concept of solid freeform fabrication, which would eventually lead to what is now known as additive manufacturing or 3D printing, was emerging for polymeric materials. Sandia National Laboratories was developing the Laser Engineered Net Shaping (LENS) process using a laser and blown metallic powder to build three-dimensional metallic parts layer by layer. These developments inspired me to submit a proposal for developing a large-scale rapid metal deposition process using an electron beam and wire feed we called "Electron Beam Freeform Fabrication" (EBF[3]). That was my first technical leadership opportunity that launched me into leading the development of this technology for 20 years for a wide variety of different applications and projects for NASA. I served in the dual role of developing the technology and continuing to advocate and support projects as a technical lead for the metallic materials area for many years.

As I continued in my career, I learned more about disciplines related to my field of materials. Materials and manufacturing processes require a knowledge of structures and systems to understand the geometries and operational requirements of components that are designed and built. Expanding my knowledge has also led to expanding responsibility supporting advanced transport technologies. I remain in a research role, but I have transitioned to more advocacy and technical leadership, generating new ideas and advocating for others' ideas to develop them into funded, executable project plans.

Epilogue

One of my most distinct memories as a young adult takes place at the Science Museum in London. I was there to see Jason, the original DH.60 Gipsy Moth aircraft in which Amy Johnson made her record solo flight from Croydon, England to Darwin, Australia.

In this pursuit, the odds were not in Amy's favor. Unlike many of her counterparts at the time, she was not given tactful public relations positioning or any kind of boost whatsoever into the world of aviation. Moreover, she was not a natural pilot and lacked natural gifts of coordination and sensitivity of touch.

Furthermore, Amy was, by all intents and purposes, an unexceptional and very average individual. She was from humble beginnings and worked as a secretary in London before stepping into an airplane (without a penny to her name) heading halfway across the world.

What made Amy remarkable was that she made the exploration flight within everyone's reach. Before she made her Australia flight, there had been very few worldwide landmarks in the history of women's flying. In fact, only four since the Wright brothers took flight in 1903. Amy would go on to be the 37th woman to receive a navigator certificate in Britain, but the world's first woman to receive a Ground Engineer License from the Air Ministry.

Amy's achievements were made possible by her relatability, determination, and tenacity, not by any excellence in flying skill. Amy's is one of the legendary stories that inspired me to pursue a career in aerospace. I thought if a girl like Amy Johnson can do it, why can't I?

It is this type of tenacity and rigor without which our great industry wouldn't be what it is today. Our tremendous progress in space discovery, aeronautical engineering, and sustainable flight would be nonexistent without the women of achievement, the thinkers, the technologists, innovators, and researchers whose stories reverberate throughout these pages.

If you look at what we have been able to accomplish—we have a lot to be proud of! Flying and space exploration have brought so many benefits to the world: Acceptance, unity, continuous learning, and discovery, all of which are critical elements for a functioning, stable society. Together we have created a movement that makes people feel like they are part of something greater than themselves. And the voices in this book make me all the more optimistic about what the future holds for this beautiful industry of ours.

Looking ahead, we have a new generation expecting us to deliver them a world that is not only brimming with technological breakthroughs but one which is better than today's. They are counting on us to continue harnessing our collective ingenuity to make progress in the name of science. They count on us to leverage innovation to better protect our planet from degradation. They count on us to better serve society, and they count on us to ensure no one is marginalized based on the unique characteristics they may have.

Most of all, they are counting on us to enable a world with an "Amy Johnson seal of approval": a world in which, even with all barriers before them and all odds against them, no horizon will be unconquerable. In this respect, I'd like to think our greatest contribution still lies ahead.

Grazia Vittadini

Grazia Vittadini is Airbus' Chief Technology Officer. Born in Italy and raised in the United States, Grazia leads Airbus' ambition to pioneer sustainable aerospace and build the future of flight.

She began her career as part of the Eurofighter Consortium before joining Airbus in Germany in 2002 where she contributed to the A380, A350, A320, and A400M programs. As she moved into more senior management positions at Airbus, Grazia was Executive Vice President and Head of Engineering of Airbus Defence and Space before becoming Airbus' Chief Technology Officer in May of 2018. As such, she is a member of the Airbus Executive Committee, the first woman to be so appointed.

Today she leads a transnational team that spans the globe with facilities in Europe, China, and the Americas that is leading the transformation to sustainable, electric, autonomous, and zero-emission air transportation.

Grazia has her private pilot's license and graduated in Aeronautical Engineering, specializing in Aerodynamics, from the Politecnico di Milano.

Grazia Vittadini

Index